TIME MATTERS

Time Matters

Global and Local Time in Asian Societies

editors

Willem van Schendel and
Henk Schulte Nordholt

BF
468
.T555
2001
West

VU University Press
Amsterdam, 2001

Editorial Note

This 21st volume of the Comparative Asian Studies series marks some changes in the appearance and organisation of the series. It is the first volume that is no longer published under the editorship of Dick Kooiman, who has stepped down as General Editor of the CAS series. We would like to thank him for his professionalism and diligence in the editing and production of the many volumes that mirror the topical and theoretical interests of Asian Studies in Amsterdam.

The present volume continues to express these interests, not only to an international public but also to our own students of Asian history and society. In this sense the publication of *Time Matters* is indicative of the increased relevance of a systematic comparison of the social and historical development of the two continents.

This edited volume is the result of the joint efforts of the Amsterdam School for Social Science Research/Centre for Asian Studies (ASSR/CASA) and the recently instituted Asian Studies in Amsterdam (AsiA) programme. It is also the first volume of the series that is published with a new cover. We felt it was time for a face-lift, and we hope the new book will stimulate people to read the volumes appearing in the series.

Leontine Visser,
General Editor
lvisser@pscw.uva.nl

VU University Press is an imprint of
VU Boekhandel/Uitgeverij bv
De Boelelaan 1105
1081 HV Amsterdam
The Netherlands
vu-universitypress.com

Editing: Vallaurie Crawford, Hawaii
Layout: Sjoukje Rienks, Amsterdam
coverphoto: John Kleinen, Amsterdam

isbn 90-5383-745-0
nugi 653

© VU University Press / CASA–ASiA, Amsterdam, 2001

All rights reserved. No part of this publication may be reproduced, stored in a retrieval system, or transmitted, in any form or by any means, mechanically, by photocopying, recording, or otherwise, without written permission of the holder of the copyright.

Contents

	List of Figures	6
	Time Matters: An Introduction *Willem van Schendel and Henk Schulte Nordholt*	7
1	The Worm and the Clock: On the Genesis of a Global Time Regime *Johan Goudsblom*	19
2	Modern Times in Bangladesh *Willem van Schendel*	37
3	Plotting Time in Bali: Articulating Plurality *Henk Schulte Nordholt*	57
4	The Quest for Universal Time: Periodising the Past in Japanese (and German) Historiography *Sebastian Conrad*	77
5	Civilising the Past: Nation and Knowledge in Thai History *Barend J. Terwiel*	97
6	Splitting Histories in Aru (Indonesia) *Patricia Spyer*	113
	References	131
	Personalia	141

List of Figures

Figure 1	Girls dressing up for the Spring Festival. Dhaka, Bangladesh, 1 Phalgun 1405 B.E. (= 14 February 1999)	44
Figure 2	Procession celebrating Bengali New Year. Dhaka, Bangladesh, 1 Boishakh 1407 B.E. (= 14 April 2000)	45
Figure 3	A Balinese calendar, 1998	59
Figure 4	One of a series of stamps – issued in 1983 to commemorate 700 years of the Thai alphabet – reproduces part of the beginning of the Rama Khamhaeng Inscription	110
Figure 5	Leaders of the revolt on the Aru Islands. Died in prison on Ambon, about 1890	119

Time Matters: An Introduction

Willem van Schendel & Henk Schulte Nordholt

On New Year's Day 2000, people all over the world watched on television how the new millennium was ushered in around the globe. For a moment, we were unified in a myth of global connectedness and technological precision.[1] Despite the fact that 1 January 2000 was just an arbitrary point in history, it seemed to acquire an ominous quality for mankind. As in the year 1000, people expected a dramatic event. In 1000 it had been the end of the world, or the end of time, in connection with the return of Christ. In 2000 it was a disastrous meltdown of the computer networks that are increasingly structuring our lives. Neither materialised. The widespread anxiety and exultation accompanying such moments point to the power that time exerts over us.

In this book we examine time in Asian history. Time is a vast subject, and it is necessary to explain our ambitions and their limits. Time matters to the contributors of this book because human beings are 'time-conscious creatures.'[2] Time is a category which expresses the human experience of duration and irreversibility, but it depends on how we constitute it. We create time as we measure it (although a great deal of time is never measured[3]), and our immersion in it is

---

1 Because the Christian era started with the year 1 instead of the year 0, the year 2000 is not the beginning of the third millennium, but the closing year of the old one. (It was only in the Middle Ages that the number zero became incorporated into Western thinking, after Europeans had imported it from South Asia via the Arab world). So the new millennium starts in 2001. Moreover, the starting point of our modern time system, supposedly the birth date of Christ, is based on sheer speculation. Since nobody knows when he was born, we may be living in 2012 or still in 1994 without knowing it, and the exact transition to the next millennium will pass (or has already passed) without anybody noticing (see Shaffer 1997).
2 Adjaye 1994: 9.
3 Östör 1993: 6.

man-made. In other words, time 'rests upon rather arbitrary social conventions,' and it is dependent on the subjective meanings which people attach to it.[4]

In this book we are not concerned with the philosophical question of what time is; we shall not treat time as a fourth dimension in which we are all immersed, or as 'a topic without any natural organising principle, one that is ubiquitous, like ether or human nature, and that, like them, baffles comparison.'[5] Rather, our focus is on how, in different historical and cultural settings, people construct time as a tool for orienting themselves in the world and organising their personal and group lives.[6] It is the changing social conventions and meanings of time which are the topic of this book.

We look especially at two relationships which are presently emerging as new topics of comparative research on time. The first is the relationship between global and local time regimes. Here we are interested in presenting different ways in which time is experienced and in showing how seemingly incompatible time regimes coexist at the local, national, and global levels. The second relationship is that between popular and academic ways of ordering time in Asian history. Here we are interested in exploring how different ways of organising and conceptualising historical time have evolved and how they shape our current understanding of time.

There are good reasons for considering both relationships within the compass of a single book. First, there are obvious links between time regimes and the ordering of historical time. One example is the tendency to break up history into the units of a time regime. When we speak of 'nineteenth-century' Thai politics, Meiji literature in Japan, or the 'Bandung generation,' we refer to time-regime units (a century, the reign of an emperor, a generation[7]) and imply that these units are also useful as historical periods because they lend coherence to historical processes occurring during them. When it is evident that there is no historical caesura at the end of a time-regime unit, many historians still cling to that unit, coming up with phrases such as 'the long nineteenth century' (which is often argued to run up to 1914) or 'the short twentieth century' (thought to come to an end in 1989). Clearly, the way we think about historical periodisation is influenced by the way in which we experience time. In this context, we may

---

4 Zerubavel 1981: xii.
5 Grew 1995: x.
6 Goudsblom 1997: 21. Gosden even goes so far as to suggest that conceptualisations of time are the most important characteristic of social formations: 'All practice creates time and the varying combinations of time within a social formation create a temporal structure or style. However, I believe that we should not merely say that social formations have their own temporal styles, but to go a step further and characterise social formations primarily in terms of their temporal styles of life.' (Gosden 1994: 187).
7 For the concept of the generation as an influential unit of periodisation in French historiography, see Nora 1997; Leduc 1999: 117-123.

speculate on how an ever-evolving 'art of timing' (Chapter 1) will continue to affect the ways in which we break up historical time into 'periods.'

Second, both time regimes and the ordering of historical time can be thought of as deriving from, and operating on, various social levels. Chapter 3 juxtaposes Balinese local time, Indonesian national time, and global time. Chapter 2 looks at three time regimes in Bangladesh, one local (Bengali time) and two global (Islamic time and 'English' time). In the same way, the ordering of historical time often differs between the local, national, and global levels, and one of the concerns of historians of Asia has been to make meaningful connections between these (Chapter 4). One point which comes up in several contributions is the constant reworking and appropriation of meanings and orderings of time between these social levels, and the fact that patterns of influence run from the global to the local as well as the other way round.

And third, both the relationship between global and local time regimes and that between popular and academic ways of ordering time in Asian history are relationships of unequal power. All contributions to this book examine the nexus of power and time in Asian societies, For example, Chapter 5 highlights the struggle between a state-imposed periodisation of Thai history and that preferred by academic historians; Chapter 6 suggests how the violence of a historical experience (the Suharto regime in Aru (Indonesia)) became codified as a rift in historical memory and led to a periodisation which sought to re-empower the Aruese as historical actors. And in Chapter 2 it is time regimes themselves which have become both tools and symbols of political confrontation in contemporary Bangladesh.

And last but not least, this book is meant as an introduction to different genres of thinking about time in the social sciences. After briefly outlining the contributions to this book in the next two sections, we conclude the Introduction with a characterisation of these genres.

Global and local time regimes

The first three chapters approach the theme largely from the vantage point of time regimes. Recent studies by, for instance, Young, Zerubavel, and Goudsblom have explored the emergence and consolidation of a single 'global time regime' as the outcome of a complex of social, economic, and technological changes in industrialising societies. These gave rise to a 'metronomic society' ruled by clocks, calendars, fixed working hours, and timetables, which then spread to envelop the entire human race.[8] In Chapter 1, entitled 'The Worm and the Clock: On the genesis of a global time regime,' Johan Goudsblom presents this argu-

8 Young 1988.

ment in terms of a theory of 'timing.' For him, timing is a mental activity based on the ability to compare processes of varying type, duration, and speed by measuring them against collectively agreed upon standards of duration and speed. Thus, the category of 'time' results from the social activity of timing. It is against this background that we must understand the emergence of a global time grid organised by the Greenwich-oriented international system of time zones.

Studies on the formation of the global time grid are important because they suggest that people all over the world eventually avail themselves of the same process of calculating time. Goudsblom argues that the time system, invented to regulate the increasing complexity of human interaction in industrialising societies, eventually comes to be experienced as a powerful outside force that we do not control and that may even terrorise us. He likens the development of this system to the habit of people attending a large gathering to raise, rather than lower, their voices in order to communicate with each other—a strategy which is counterproductive because the collective noise they produce makes communication more difficult. Like such people, we feel that we are the victims of a system that we have created ourselves. We experience the time regime, which was designed to help us regulate our lives, as an implacable outside force constraining even the most intimate aspects of our lives.

Useful as such insights may be, studies on the formation of the global time regime also raise a number of doubts. They have been criticised for remaining Eurocentric and for not dealing with other time regimes in different cultures.[9] Because studies of global time are convergence-oriented, as such, they are eager to highlight global unity, and are less impressed by and informed about local difference, which is hastily viewed as pre-modern, non-progressive, and as being gradually marginalised.[10] These studies provide no answers as to how the new global time regime has been contested, and whether it has indeed been as successful as they claim in reorganising human life world-wide. Among the questions that they leave unanswered are, for example, how did the global time regime actually mesh with numerous local time regimes all over the world? What were the conditions in non-industrialised societies that allowed global time to be integrated in, or perhaps appropriated by, local times? How did local times mould the present form of the global time regime? Which forms of resistance to global time could, and can even now, be identified?

In this book, we do not pretend to provide coherent answers to all these questions. Our main purpose is to put them firmly on the academic agenda. They all

---

9 For civilisational approaches to time in Asia, see Nakamura 1968 and Needham 1968. Recent time studies in the historical-anthropological mode include Adjaye (ed.) 1994, Hughes and Trautmann (eds.) 1995, Munn 1992, Östör 1993, and Hoskins 1993. For studies on the pace of life, see Levine 1997.
10 For example, Adjaye speaks of 'either outright denial or distortion by Western authors' of the subject of time among Black peoples in Africa and the diaspora.

relate to the central concern of this book: to demonstrate that the world-wide social activity of timing (to use Goudsblom's phrase) is much more complex and varied than existing studies on the global time regime suggest. The convergence-oriented approach emphasises the gradual rise of a global uniformity in time measurement, a standardisation of the units of clock time and a co-ordination of calendars. By implication, local time regimes are being marginalised and, at best, end up in the domain of subsidised folklore. This approach is unable to conceptualise the co-existence of, and interaction between, different time regimes. As a result, it is insensitive to the plurality of time regimes in which most people live.

The convergence-oriented approach stands in a long tradition which privileges Western time as modern time. Europeans who travelled beyond the confines of their continent as administrators, merchants, missionaries, or scholars were often unable to acknowledge that the 'natives' they encountered in Asia, Africa, and Latin America were living in the same age as they themselves. Instead, they categorised these non-Europeans as belonging to an earlier stage of development which somehow had managed to survive in the interstices of the modern world. Europeans could descend into this earlier stage, after which they could re-enter their own advanced age.[11] The same evolutionary idea lies at the basis of the well-known dichotomy between 'tradition' and 'modernisation' and reflects the analytical poverty of large parts of the social sciences in the 1950s and 1960s. The distinction between the modern and the traditional was predicated upon the assumption of a dynamic Western world characterised by linear development juxtaposed to a stagnant non-Western world caught in cyclical notions of time. Acceptance of the Western time regime and the behaviours that it implies would pave the way to progress and welfare. Non-Western notions of time were considered to be traditional and largely informed by religious beliefs which prevented development—despite its Christian pedigree, the Western time regime was considered to have become secular. We summarise these contrasts as follows.

Western time	*Non-Western time*
Modern	Traditional
Secular	Religious
Linear	Cyclical

---

[11] Fabian 1983.

Since things could not be *both* modern *and* traditional, *both* secular *and* religious, or *both* linear *and* cyclical, it followed that there could be only two opposite time regimes: the global/Western one and the local/non-Western one. The former was linked with development, the latter with stagnation.

Nowadays, however, there is very little support for such polarised worldviews. It is no longer possible to distinguish the West from the non-West in any productive, meaningful, or precise way. There is as much 'non-West' in the 'West' as the other way around. San Franciscans celebrate Chinese New Year, residents of Rotterdam join the Caribbean Carnaval, Buddhists in Bangkok decorate their houses for Christmas and African-Americans invent the Kwanzaa festival. Clearly, the modern and the traditional as mutually exclusive categories have become hopelessly outdated. Social scientists have realised that social phenomena are best understood without such *a priori* labelling. When studying time regimes, we need to concentrate on the articulations between such regimes and we should realise that old time regimes are quite capable of being modernised to suit new needs.

We are only beginning to explore how non-European and European concepts of time have interacted. On the one hand, there is the immensity of Indian 'Big Time,'[12] or the international Islamic time regime, and, on the other hand, there are the 'small times' of specific localities.[13] Little is known about how these have been transformed, reformulated, marginalised, revitalised, or, perhaps, abandoned. What role did concepts of time play in both the formation of colonial states and new forms of economic exploitation, as well as in struggles against colonialism and capitalism? How were they employed, and resisted, in bids for self-determination and nation-building? In many Asian societies, various measurements of time continue to co-exist. Why is this, and what do such relatively rich temporal repertoires reveal about the social groups that use them?

These questions are addressed here by the provision of examples of how various groups of people in different Asian societies have constructed time systems over the last one and a half centuries. In Chapter 2, entitled 'Modern Times in Bangladesh,' Willem van Schendel looks at one particular Asian society, which can be called time-rich because it applies three different ways of measuring time. These are continually revitalised by identity politics, which stress the connection between self, time, and the future of society. In Bangladesh, how a person measures time symbolises that person's political ideas. The main conflict is symbolised by the use of Bengali and Islamic time. The use of the Bengali calendar first became a symbol of the political aspirations of the Bengali regional elite in East Pakistan (as Bangladesh was known till 1971) and is now associated with ethnic assertion, religious tolerance, and progressive politics. The use of the Islamic

---

12 Trautmann 1995.
13 Cohn 1995.

calendar has become associated with religious assertion, conservatism, and a moral revival. The global time regime (known as English time in Bangladesh) is of much less account. Disliked by many as a Western and Christian import but admired by others as a symbol of modernity and global 'cool', it is used as a politically neutral system. In Bangladesh, as elsewhere, concepts of global time are domesticated and adapted to local needs. Local and global concepts of time are continually manipulated and reinvented to serve new political needs, and to give expression to social identities. As long as multi-temporality serves such purposes, it will flourish.

In Chapter 3, entitled 'Plotting Time in Bali: Articulating Plurality,' Henk Schulte Nordholt shows that several time systems in Bali not only exist side by side, but also interact. Based on several case studies, he argues that there is 'a complex articulation between a state cum capitalist time regime and a local ritual and calendar system,' which allows people in Bali to communicate and realise strategically important (regional) counter-identities. Significantly, the local ritual calendar is not static, but has been and still is being modernised. The reproduction of the ritual time system is the result of a complex interaction between institutions at the level of the state, and regional agents who define the Hindu identity of the Balinese, as well as its profitability, in terms of tourism. The identities which are strengthened by the use of the Balinese ritual calendar are supported by specific ideas of how the past and the present are related, and these are quite different from the ideas held by most Western-trained historians.

Academic and Popular Ways of Ordering Time

The three concluding chapters of this book focus on contests regarding the ordering of historical time. If the past is an unstructured chaos, history is the attempt to make sense of it by ordering parts of it in a meaningful way.[14] Time has been a crucial tool for historians in creating narratives of the past: by means of a linear chronology, they have sought to describe processes and establish causal relationships; by means of periodisation, they have sought to identify coherent stretches of 'continuous time' separated by disruptions and crises.

The usefulness of this enterprise has always been debated at various levels. Is historical narrative determined by its own plot rather than representative of any 'real' history? Can historical narratives be improved by juxtaposing 'clock time' and 'subjective time'?[15] Is the linear concept of time part of a 'bourgeois narratology,' and should 'chronology itself, the sacred cow of historiography, [...] be sacrificed at the altar of a capricious, quasi-Puranic time which is not ashamed

---

14 Huizinga 1946: 9.
15 Cohen 1968.

of its cyclicity?'[16] Does historical scholarship take sufficient note of the different ways in which human groups have valued time?[17] These questions go well beyond making the usual connection between industrial capitalism and timing—i.e. the realisation that 'time is money' and that its precise measurement is essential—and embrace the fact that social groups have assigned different *valuations* to their past, present, and future.

In Chapter 4, Sebastian Conrad explains how this debate dominated Japanese history writing in the second half of the twentieth century. His contribution, entitled 'The Quest for Universal Time: Periodising the Past in Japanese (and German) Historiography,' is concerned with the emergence of a new mode of ordering historical time in Japan after 1945. He argues that it is crucially important to understand the ideological underpinnings of periodisation in Japanese accounts of the past. The 'poetics of chronology,' or the persuasiveness of the historian's account, rests on the interpretative force that he or she gives to the constructed discontinuities between, and the coherence within, different periods. In the case of Japan, Conrad provides us with an example of an Asian society struggling with an academic historical tradition which was predicated upon periods derived from the European past (antiquity, the Middle Ages, and so on), and which focused on the description of a 'nation' and its 'national' space. After Japan's defeat in World War II, a group of influential historians rejected the narrowly nationalist historiography, which had legitimated Japan's expansionism, and searched for a more objective and 'scientific' approach. Their quest for universal time led them to embrace Marxist periodisation and brought about a 'chronometric turn' in Japanese thinking about the past.

This conflict between particular political interests and academic concerns is also the dominant theme in Chapter 5, written by Baas Terwiel. A growing interest in popular cultures and their ways of dealing with history is currently sharpening these debates all over Asia. Official historians create significance and coherence with the support of recorded memories and documentation, which are often related to the interests of the nation-state. Popular historians who emphasise a local, non-state orientation, in Asia as elsewhere, use very different 'sources' to order time and create significance. As competing guides to authority and action in the 'real world,' the various concepts of time they use matter enormously. Are the academic conventions regarding the measurement and meaning of time among official historians of Asia being marginalised as popular versions of history come to the fore? Should historians confront these popular histories with their own and relatively autonomous notions of time as new forms of myth-making in the service of current political and economic interests?

16 Guha, 1996: 12.
17 Leduc 1999: 91-133.

In 'Civilising the Past: Nation and Knowledge in Thai History', Terwiel describes the acute tension between a state-sponsored version of Thai history and the past as understood by academic historical researchers. Much of the periodisation of an idealised Thai past, which can be found in officially approved textbooks and school curricula in Thailand, has been proved wrong by historical research based on primary sources. Yet it has survived without much explicit opposition. Terwiel explains this anomaly by pointing to the authority ascribed to certain kinds of knowledge. There is a strong tradition in Thailand which sees teachers as the transmitters of a well-nigh perfect body of knowledge and which does not invite personal inquiry, let alone critical commentary. This tradition, and 'the government's active role in using knowledge for its own nationalistic purposes,' has made it very difficult for Thai historians to challenge the state-sponsored version of the past. Clearly, time matters enormously in Thai history writing. Thai historians who challenge the officially imposed nationalist periodisation based on the authority of primary sources expose themselves to the charge of disloyalty to the state.

The concluding chapter of the book, 'Splitting Historiographies,' takes us to Aru, a group of islands in eastern Indonesia. By means of three case studies, Patricia Spyer shows us how the Aruese value two large categories of time: the past and the present. She is particularly interested in the break, or 'split,' which is constructed between the past and the present, and how this split is symbolised in ritual performances, such as songs, conversions, or seasonal rites. Here, local forms of history-making rework the national historiography of Indonesia, giving voice to local ideas of time and space and endowing them with authority. Women and men in Aru assemble and codify 'tokens of nationalist historiography which tend to come to the island as broken-off bits of a more monumentalised state project.' In this process they are indebted to official versions of national history and to that extent consolidate their own subordinate position within the hierarchy of the nation. But at the same time they are the creators of a historical knowledge which is recognisably Aruese and which establishes their own agency within the nation.

All three chapters explore the problems of attuning the ways in which historical time has been ordered at different levels of society. Conrad explains the Japanese historians' quest to inscribe Japanese 'national' history into world history. Terwiel highlights the confrontation between the Thai state elite's unchanging version of 'national' history and a research-based vision of a multi-interpretable past which needs a different periodisation. And Spyer shows how 'national' and 'regional' historiographies intersect without necessarily merging.

Genres of Thinking About Time

The contributions to this book look differently at the ways in which time matters in Asian histories. This is not only because our examples come from societies in East, Southeast, and South Asia, or because they deal both with how time is experienced and how it is organised. Equally important is that the book brings together perspectives from different social science disciplines and can serve as an introduction to several styles of studying time.

We can arrange the contributions in a grid based on two dimensions: the discipline which has most inspired them and the level of society on which they focus. Thus Goudsblom, Van Schendel and Schulte Nordholt provide examples of how time is thought of in historical sociology on the global, national, and local scale, respectively. Conrad's contribution stands in the tradition of historiography, the critical analysis of how professional historians have conceptualised historical time, and his interest is focussed at the national level. Terwiel, on the other hand, applies the specific perspective of cultural analysis which has developed within area studies, and which owes as much to conceptualisations of time in the humanities as in the social sciences. His chapter privileges the national level. And Spyer approaches the theme from an anthropological background, stressing the symbolic uses of time by local actors.

In this way, the chapters of the book point to important genres of interpreting time. There is still remarkably little connection between these genres, which have developed among largely separate groups of scholars in different university departments and publishing in journals geared at their own disciplinary colleagues. But as we have attempted to prove in this book, these genres need to be juxtaposed and integrated in order to deepen our understanding of the remarkable variety of ways in which time has been, and continues to be, experienced, ordered, and contested.

Finally, all contributions to this book deal with the relationship between time and social power. It is a central concept in Goudsblom's contribution, which posits the power of an expanding time regime, to which people all over the world can do little more than adapt. Van Schendel highlights that assumptions of global time convergence need to be qualified. He shows how different ways of measuring time came to play a symbolic role in national politics in Bangladesh, thereby opening up a local arena in which global time was held at bay. Schulte Nordholt and Spyer present cases in which time is similarly used in the creation and maintenance of regional identities in Indonesia, against a national time regime and a state-centred national historiography. Terwiel explores the rigidities of nationalist views of time in Thailand and the restricted opportunities for challenging them within that country. And Conrad shows how power mattered in thinking about time in Japan: It was the shattering of Japan's dreams of empire in 1945 which opened a window for reworking dominant views of time in Japanese historiography. It led to an attempt to replace nationalist periodisation with an approach which would anchor Japan's past in a universal or world history.

The technology of timing has reached an unprecedented level of accuracy: it can measure a millionth of a second as easily as thousands of light years. In view of such prowess, it is important to reflect on the fact that still a great deal of time is never measured. For example, individual memory is able to escape the rigours of modern clock time by organising time in a relatively autonomous way. When memories are represented in the form of stories or oral history, this autonomy can become an object of study. It is only then that we realise how much timing is in fact an individual domain, interconnected with the time regimes that prevail in the individual's society but not completely predetermined by them.

Our main interest in making this book, however, has not been to draw attention to those interconnections but to the social and political contexts in which time is constructed. Time should matter to historians and other social scientists because available time-regime studies assume a global sway but tell us far too little about the local uses to which time regimes are put. We argue that, first, the global time regime is domesticated differently in different parts of the world, and, second, that most people live in a plurality of time which is the outcome of a historical interaction, or articulation, between various time regimes. We also argue that exercising power implies controlling and imposing time, and that we may therefore analyse struggles over time and multitemporality in terms of particular constellations of power. Finally, the essays in this book wish to demonstrate that our conceptions of time shape our views of the past and thereby our views of the present and how that present may be turned into the future. As such, the study of the construction of both time regimes and historical periods needs to become an integral part of social inquiry.

1 The Worm and the Clock: On the Genesis of a Global Time Regime[1]

Johan Goudsblom

> Time is money. Haste is debt.
> J.A. Emmens

One of the most remarkable aspects of globalisation is the spread of a uniform system of time measurement around the world. Nowadays, anyone with an accurate watch can tell to the second what the time is for nearly any location worldwide. If we know, for example, that it is forty-three minutes after 12 o'clock in Amsterdam, we can say with absolute certainty that it is forty-three minutes past the hour in Tokyo or Rio de Janeiro—while also knowing that in Bombay it is thirteen minutes past the hour, since the time difference in India is thirty instead of sixty minutes.

This all refers to the division of time for a single day. With regard to years, months, or weeks, there is less uniformity. Variations occur when certain religions or nations use specific calendars with particular arrangements of months and weeks. However, the number of such cases is small. By far, the most commonly used calendar is the Christian calendar. It is used in China as well as in some Islamic countries. In Japan, two calendars are officially recognised: the Christian and the imperial, though in daily practice, people primarily hold to the Christian calendar.

Usually, the world-wide system of time measurement is left out of the current literature on globalisation—most likely because it seems to cause relatively few problems. This last point, however, can also be an incentive to examine such a

---

[1] Originally published in Goudsblom 1997: 20-38. I wish to thank Hes Godschalk-Hessenaur, Jona Oberski, Fred Spier, and Nico Wilterdink for their useful commentary.

notion somewhat further. This is particularly so given that the very procedure of time measurement is essentially *locally* bound. How can it be, then, that a single uniform system of time measurement has emerged for the entire world?

The Concept of Time

In daily life, we rarely stop and wonder about the concept of time and, if we do so, we usually tend to perceive time as a 'natural, ahistorical, and unproblematic' given.[2] Upon closer examination, however, such characterisations are rash. Many discussions of the nature of time begin wisely with the words of St. Augustine who, commenting on the meaning of the concept of time in his *Confessions*, remarked: 'Provided that no one asks me, I know. If I want to explain to an inquirer, I do not know.'[3]

The concept of time can be defined in a number of ways. One way is to look at it as being inexorably bound to individual experience, whether that be as a purely subjective inner sensation or as a general category of perception in the sense of Kant. On the other hand, there is the view that time is a natural process which carries on independently from human life; even if there were no humans present to experience or observe time, time would still march on. That is, the earth keeps turning and the 'fourth dimension' continues to exist, regardless of any individual person.

The gap between these two views appears unbridgeable. Perhaps, however, we can bring them closer together if we take a third, more sociological approach. We may then regard time as a socio-cultural construction which aids people in their efforts to collectively orient themselves in the world and to co-ordinate their actions.[4] This view represents the position taken by Norbert Elias.[5]

'Timing,' according to his theory, is a mental activity based on the ability to compare processes of varying type, varying duration, and varying speed by measuring them against standardised processes of collectively calibrated duration and speed. Thus, processes which take place inside the human body (such as blood circulation or reading one's pulse), social processes (such as a game or a meeting), and processes in the outside world that carry on completely independently of people (such as the revolution of the moon) can be seen and understood as belonging to the same 'dimension.'

For us today, it has become nearly impossible to describe the experience of time without referring to the institutionalised forms of measurement—the in-

---

2 Rotenberg 1992: 2.
3 Augustine 1991: xi, xiv, 17.
4 Good overviews of the sociological theories about time can be found in Adam 1990, Nowotny 1994, and Schmied 1985.
5 Elias 1992.

struments, the techniques, and the terms—which have been developed in human society to provide a common denominator to all those manifold processes. The same institutionalised forms also provide the basis for all scientific knowledge about time as a process of nature.

However plausible this sociological vision of time may be, it is certainly not self-evident, nor generally accepted. To illustrate this point is a quote from Jeremy Campbell who, in a book on the very subject of time, simply considers time as a phenomenon which exists on its own and, as such, seems to require no further elucidation:

> In the simplest sense, time is important for most living species. As time goes on, the environment changes in a periodic fashion, and the changes have meaning for the organism. An animal's whole way of life, indeed its life history as an individual, may be designed to take advantage of changes that are beneficial and to avoid changes that are harmful. Day alternates with night, and the environment of day is very different from the environment of night. In temperate zones the seasons follow their annual cycle, from warm to cold and back to warm again, as predictably as a clock. A simple type of animal, lacking any biological means of opposing these changes, would have to adjust its way of life accordingly. Time would be a dictator, a tyrant. The animal would be cold when the weather is cold, and warm [...] when the weather is warm. An earthworm's temperature is the same as that of the soil in which it moves, and a fish is neither warmer nor cooler than the sea in which it swims. Such 'cold-blooded' animals, which are not necessarily cold but take on the temperature of their surroundings, are at the mercy of the clock and the calendar. Only if the physical environment were to remain the same 24 hours a day, 12 months a year, would such an animal's behaviour be unaffected by time.[6]

This is a fluently written passage, the tenor of which—at first glance—cannot be disputed. The facts to which it refers are irrefutable. There is, however, something profoundly wrong in the way these facts are rendered. To maintain that an earthworm is tyrannised by time displays a fundamental misunderstanding of the meaning of the concept of time.

The earthworm is indeed extremely susceptible to the natural conditions in which it finds itself. The 'internal rhythms' of the organism adjust automatically to the 'external rhythms' of its environment; changes in the outside temperature bring about changes in the inside temperature of its body. This in no way means, however, that the earthworm is obeying a clock or calendar. On the contrary, the earthworm has absolutely no contact with anything remotely resembling a clock or calendar. The only way for the earthworm to be 'at the mercy of

---

6 Campbell 1986: 43.

the clock' would be if it were transferred into a laboratory where humans could raise or lower the temperature by a few degrees according to a fixed time schedule. Then, and only then, would the worm live under a time regime.

What a Dutch proverb has to say about sociability applies also to the earthworm: it does not know time. Worms can only respond directly to what is happening around them. Only humans are aware of time. To help them get a firmer grip on the manifold processes involving and surrounding them, humans have developed the means for comparing the duration and speed of all these processes to each other.

This, in brief, is the sociological view of time developed by Norbert Elias. One can only speak of time if—in addition to individual experiences and observations, as well as processes of nature which take place independently of people—the third level is also taken into account: the level on which people 'use a socially standardised sequence in order to compare sequences that are not directly comparable.'[7] The category of 'time' results from the social activity of 'timing.'

The Development of Timing

Roughly four phases can be distinguished in the development of timing that have led to current practice. I will first give a short characterisation of each and will then elaborate on them successively. To distinguish each phase, I will refer to the particular instruments developed for marking the consecutive segments of the most elementary unit of time, the day. In *phase one*, there were no such instruments and, as a result, it was not possible to divide the day into clear-cut intervals. *Phase two* is marked by the development of different types of instruments such as the sundial, the water clock, and the hourglass, with which people could measure time in the localities in which they happened to find themselves. During this phase, a system of dividing the day into twenty-four hours emerged; it remained, however, that not only did the hours vary in length from place to place, but, in most places, the length of the hours would also change considerably from day to day in the course of the year. *Phase three* is distinguished by the development of mechanical clocks which, with fixed hours of equal length, made standardised time measurement possible; matching the clocks to each other also became possible, at least at a local level. *Phase four* represents the emergence of new techniques ranging from faster means of transport and telecommunication to the implementation of a global system of standard time, thereby creating the conditions to synchronise time measurement to the tiniest fraction of a second anywhere in the world.

7 Elias 1992: 2.

Obviously, the enumeration of phases does not say anything about the actual chronological order in any specific case. The transition from phase one to phase two, for instance, could have occurred in different societies at very different moments. Moreover, one or two phases can be skipped. This has been the case during the twentieth century, as societies which previously had no time-measuring instruments were at once exposed to the generally applied grid of globally standardised time.

Furthermore, the implementation of a later phase of timing in any particular society does not mean that traces of earlier phases are automatically and totally erased. Even in societies with a social rhythm which is strongly regulated by the clock, there are still many opportunities for people to withdraw from the time regime. The margins for doing so, however, are limited and one can only 'forget time' temporarily. From its first day, a baby born in the contemporary Netherlands will grow up in a social world drenched in time consciousness. Rhythms of sleeping and eating are immediately fitted into a schedule of hours.[8] While the degree to which people are subjected to the regime of the clock varies around the world—both within and between societies—the regime as such has penetrated everywhere.

The time grid is not limited to days and hours, but extends to longer periods such as weeks, months, years, centuries, and aeons. Calendars and eras, marking longer periods of time, will also be referred to below. I will focus, however, upon the development of hours as the unit of time measurement during the day and on the subsequent division of hours into smaller units.[9]

Phase One: No Time Measurement and No Hours

Phase one represents the period extending the furthest back in human history, and I will deal with it only shortly here. In the absence of instruments for measuring time, people had to rely on signals derived directly from natural processes. Human beings find themselves continuously confronted with such 'natural signals,' originating both outside and inside their own bodies. Any decision to act depends on their interpretation of those signals. Thus, people may decide to go to sleep because they either feel tired or because it is getting dark outside; in either case, the decision is based upon a signal from a natural process and not upon a specially designed instrument.

A third type of indication that the moment has come for a particular activity is given by the rhythm of social life. In the account of his research of the Nuer People in East Africa in 1930 and subsequent years, the anthropologist E.E.

---

8 Cf. Gleichmann 1983.
9 In the interest of space, I am leaving out the development of the week. See Zerubavel 1985.

Evans-Pritchard has provided a graphic description of how the Nuer spent their days according to a common rhythm which was primarily determined by the sequence of events associated with tending to the cattle:

> The daily timepiece is the cattle clock, the round of pastoral tasks, and the time of day and the passage of time through a day are to a Nuer primarily the succession of these tasks and their relations to one another. The better demarcated points are taking of the cattle from byre to kraal, milking, driving of the adult herd to pasture, milking of the goats and sheep, driving of the flocks and calves to pasture, cleaning of byre and kraal, bringing home of the flocks and calves, the return of the adult herd, the evening milking, and the enclosure of the beasts in the byres. Nuer generally use such points of activity, rather than concrete points in the movement of the sun across the heavens, to co-ordinate events. Thus a man says, 'I shall return at milking,' 'I shall start off when the calves come home,' and so forth.[10]

The Nuer people did not use any special instruments to measure 'time,' nor did the concept of time exist in their language. The only specific units of time known to them were the day, determined by the cycle of lightness to darkness; the month, determined by the circulation of the moon; and the year. The ever-returning annual succession of wet and dry seasons had a profound effect upon their lives, and they indicated the seasons with words principally referring to the pertinent seasonal activities. They had no need for either clock or calendar, and when Evans-Pritchard refers to the cattle as their clock, it is clearly metaphorical.

Phase Two: Variable Time Measurements, Unequal Hours

Clearly, in phase two, people did not cease to rely on natural signals. Rather, in addition to the previous methods, special artefacts were designed for the purpose of timing. It has been suggested that megalithic constructions such as Stonehenge might have filled this function.[11] Only from later periods, however, are instruments found which were clearly intended to indicate time as a primary function. The most well-known of these are the sundial, which registered the movement of the sun by capturing the shadows it created; the water clock; and the hourglass, which made use of a mechanical process.

Townspeople are to have used sundials primarily as a clock: the location of the shadow indicated how much of the day had passed. However, because the position of the sun changes from day to day in most places on earth, people were

---

10 Evans-Pritchard 1940: 100-101.
11 Cf. Aveni 1990: 74-80.

also able to tell how far advanced the year was from the length of the shadow, and how far away they were from the shortest or longest days of the year. In this way, the sundial also served as a calendar.

While this second function of the sundial was slightly more complicated, it was probably more useful to agricultural societies. For farmers, the division of the day was hardly a problem. Whether it was the season for sowing or harvesting, once they began their work they had little need of instruments to tell them the time; the division of the day was based on the regularly recurring activities of working, eating, and resting. The position of the sun and their own sense of hunger or fatigue were enough for them to assess how far the day had progressed.

Of much greater interest to the farmers was the question regarding the best time of year for sowing or harvesting—a question which was often turned into a matter for experts in establishing the calendar. Their opinion could also be crucial for other important decisions, such as when to carry out a military campaign. In many early agrarian societies, a class of priests arose who took on this special responsibility of determining when the day had come for particular collective activities; the close relationship which resulted between religion and the calendar still persists almost everywhere.[12]

In cities, where there was no longer one tacitly accepted common rhythm ruling daily life, the need emerged to have, in addition to a calendar, a more sharply defined division of the day. Sundials fulfilled this need by making it possible to observe distinct parts of the day on a two-dimensional surface. It is quite likely that the invention of the sundial was a necessary condition for the invention of the *hour*: one-twelfth of the distance covered by the shadow of a sundial in the course of a day. The figure twelve corresponded to the favoured duodecimal counting system of the Babylonians, who are known as the first to have implemented this method of timing. By adding twelve additional unobserved nighttime hours to the number of measured daytime hours, they established the twenty-four-hour time framework for one solar day.

Dividing up the solar day (also referred to simply as 'day,' creating some confusion) into twenty-four hours was an important step in the standardisation of timing. This standardisation, however, remained at a very local level: the sun always appeared earlier above Babylon than above the more westerly situated Damascus; determining the time differential in hours between the two cities was practically impossible.

An additional complication arose out of the north-south differential. During summer days, the sun stayed longer above Damascus than it did above the more southern Jerusalem. As a result, the hours were longer in the summer in Damas-

---

12 For more reading about the method of timing by priests and the power they derived from it, see Goudsblom, Jones, and Mennel 1996.

cus, with the reverse being true in the winter. The further the distance from the equator, the greater the differences. In the Netherlands, at a northern latitude of fifty-two degrees, the sun stays twice as long above the horizon during the summer as it does during the winter. Therefore, in the Middle Ages, when the Babylonian system of measuring hours was still in use in the Netherlands, an hour in the daytime in June was twice as long as an hour at night.

In response to the drawbacks of hours of varying lengths associated with the sundial, first the water clock and later the hourglass could offer a remedy. These instruments operated independently of sunlight—allowing their use during the night as well as during cloudy skies—and always measured, at least in principle, the same span of time. There were also some practical disadvantages, however. Water clocks and hourglasses were only capable of measuring limited spans of time; and the time span varied from device to device. Moreover, both types of instruments were strongly affected by the weather: during a frost, the water clocks would freeze, and humidity would cause the hourglasses to clog up.

When sundials were put to use in the third century BC in the city of Rome, satirists cursed the tyranny to which people were now subjected. Mealtime, they complained, was no longer to be determined by one's stomach, but by the sundial. Still, the discrepancies between all the various sundials and water clocks were very large in comparison to the strict standards of our modern time regime. In the first century AD, Seneca lamented that people were more likely to reach consensus about problems of philosophy than about the time of day.[13]

Until early modern times, people had to put up, just like Seneca, with all sorts of time measurers which, each in their own way, divided the day into hours of varying lengths. In his book about feudal society in Europe, Marc Bloch described the situation as follows:

> These men, subjected both externally and internally to so many ungovernable forces, lived in a world in which the passage of time escaped their grasp all the more because they were so ill-equipped to measure it. Water-clocks, which were costly and cumbersome, were very rare. Hourglasses were little used. The inadequacy of sundials, especially under skies quickly clouded over, was notorious. [...] Reckoning ordinarily—after the example of Antiquity—twelve hours of day and twelve of night, whatever the season, people of the highest education became used to seeing each of these fractions, taken one by one, grow and diminish incessantly, according to the annual revolution of the sun. This was to continue till the moment when—towards the beginning of the fourteenth century—counter-poise clocks brought with them at last, not only the mechanisation of the instrument, but so to speak, of time itself.[14]

---

13 Boorstin 1991: 50.
14 Bloch 1961-1962: 73-74.

Bloch did not even mention the problems with the calendar in this passage. The yearly course of the earth around the sun amounts to 365.2242 days and cannot possibly be expressed in an exact round number. The movement of the moon around the earth—measured by the amount of time between two successive new moons (the so-called synodical month)—lasts 29.5306 days, merely adding to the confusion. Egyptian astronomers in antiquity attempted to make years, months, and days correspond better to each other by rendering the passage of months independent from the position of the moon, and by periodically adding an extra day to the year. During the time of the Roman Empire, Julius Caesar used this information and experience to implement a calendar made up of months of unequal lengths as well as a leap year every four years.

The Julian calendar was, in turn, taken up by Christian Europe. However, it continued to display certain shortcomings which eventually led to a discrepancy of more than ten days between the official calendar time and the time according to the position of the sun. In response to this, Pope Gregorius XIII introduced some further refinements to the calendar in 1582 which, by making the leap year system slightly more complicated, improved the synchronisation. The Gregorian reforms were first accepted by the Roman Catholic countries, and later by the Protestant countries of Europe as well; with a few minor adjustments, the same system is still used today. On the other hand, the history of China—with more than fifty reforms made to the calendar by imperial decree over twenty centuries—shows far less continuity in this respect.[15]

With the emergence of the diverse devices for timing, a general concept of time also developed, suggesting a synthesis of the various forms of measuring and experiencing time. The words of St. Augustine cited above illustrate the puzzlement aroused by reflecting upon the elusive yet inescapable character of 'time.'

Phase Three: Standardised Time Measurement and Equal Hours

Just as the sundial created the technical conditions necessary for dividing the day into twenty-four hours (albeit variable), so did the invention of the mechanical clock enable people to standardise the hours into time units of equal length everywhere. This invention took place in Western Europe around 1300. In the beginning, the 'hour-clocks,' driven and regulated by weights and counter-weights, did just that: they indicated the hours or, in a few cases, also the quarters of the hour. The large mechanical clocks of the Middle Ages were unfit for measuring shorter spans of time, which continued to be gauged by sundials and hourglasses. Yet, from the very beginning, the clocks had one great advantage

---

15 Macey (ed.) 1994: 73.

because, in principle, they could function continuously without interruption, day and night the whole year through. Provided that the weights were monitored and the mechanism was adjusted as necessary, the clock rang out the hours with a set regularity.

Jacques LeGoff described how, as early as the fourteenth century, the standardised hour of the city hall and stock exchange towers in the cities of Western Europe overtook the more variable time units into which the church divided the day according to the moments of prayer. Everywhere, churches were forced to capitulate and adjust to the division of the day as determined by the mechanical clock. Before long, the hours also rang from the church towers, as well as from the city halls and market halls.[16]

A new time regime began with the implementation of mechanical clock time —one that was tighter and stricter, more secularised and more rationalised. If the residents of old Rome were already complaining about the tyranny of the sundial, the mechanical clock in medieval cities made the division of the day even more rigid. The implementation of standardised hours created the conditions for the type of labour relations which Karl Marx later typified as being specific to capitalism: relations in which the owners of the means of production employed others to carry out the actual labour, who were then paid not according to their performance, but according to the number of hours they had worked as wage-earners. Piece-wages became hourly wages, and time became money.

In Chapter 8 of the first volume of *Capital*, Marx described the struggle carried on since the fourteenth century between capitalists and workers over the length of the workday.[17] In doing so, however, he passed over the preceding conflict: before the struggle over the number of hours in a working day was the conflict about the length of the hours. Initially, workers were in favour of mechanical clock time because they saw it as their only protection against manipulation of the church hours carried out by the owners to their own advantage. Soon enough, however, the owners figured out ways to profit from standardised hours for workers. As such, both parties resigned themselves to the standardisation of hours, and replaced the struggle over the length of the hour with the struggle over the number of working hours.[18]

By Marx's time, the system of dividing the day into standardised hours and minutes was beyond dispute. Marx's own theory of capitalism was entirely predicated upon the existence of the 'hour' as an inflexible unit of account. Indeed, we could hardly imagine a form of capitalism without the clock and calendar.

---

16 LeGoff 1977: 46-65. Dohrn-van Rossum modifies LeGoff's view by indicating that the municipal governments attached their own level of prestige to the public clocks. See Dohrn-van Rossum 1996: 140-150.
17 Marx 1978.
18 Cf. LeGoff 1977: 66-79. See also Thompson 1967.

The control and payment of workers, as well as the transactions between entrepreneurs, are based upon a generally accepted form of time measurement.

Phase Four: Synchronising the Hours

What had not yet taken place in the time of Marx was the synchronisation of the hours. Indeed, by the middle of the nineteenth century, the clock time of Europe —with a standardised division of the day into twenty-four hours, and each hour having the same length of sixty minutes—had been implemented in many parts of the world. Numerous clocks of good quality, off by just a few seconds per day, could be found. In most places, clocks were collectively set based on one common time. The common time, however, differed from place to place.

For a long time, this difference did not cause difficulties. What did it matter that the clocks in Amsterdam ran a minute or so ahead of those in Haarlem? Not until the development of rapid transport and means of communication did this lack in synchronisation actually become a problem. At the same time, though, that the modern transport and communications systems highlighted the difficulties resulting from a lack in synchronisation, they also facilitated the solution. The more people travelled, the more opportunity they had to compare the local times; along with this, the telegraph made almost simultaneous comparison of times between even larger distances possible.

Wherever there were train stations and telegraph offices, clocks were co-ordinated at a national level. England led the way, as it did in so many respects, and other countries followed. Everywhere, protests were heard from local notables who felt that a piece of local identity would be lost with local time. Modernisation, however, was not to be stopped; the mosaics of local times made room for the unified grid of a single national time. Just as the time of the stock exchange had triumphed over the time of the church, now the triumph came for the time of the station.

Around the end of the eighteenth century in England, a standard national time was slowly emerging, mainly as part of the attempt to increase the efficiency of stagecoaches and postal services. The rapid emergence of the railways after 1830 gave extra impulse to this development, and helped to relieve the chaos of uncoordinated service times and timetables, which often caused many inconveniences and also increased the likelihood of collision.[19] Almost as a matter of course, the establishment of one national standard time was based on the time in London. Not only was it the capital city, but it was also the terminal point of many railway lines, as well as the seat of the astronomical observatory at Greenwich, specialised in the practice of timing ever since its establishment in 1675.

19 Zerubavel 1982: 9.

As already said, the British example was followed by other countries, despite some protests. Within a few generations, standard time was realised everywhere. In connection with the theme of globalisation, it is worth noting that international time co-ordination has almost immediately followed national co-ordination, and in some respects, has even preceded it.

The Emergence of a Global Time Grid

The question of when and where the system of a single standard time for the whole world took its shape can be answered with great precision: it was in October 1884 at the International Meridian Conference in Washington, DC. Representatives from twenty-five countries decided then and there that the world would be divided into twenty-four time zones, each with an east-west span of approximately fifteen degrees. The prime meridian would run through the observatory in Greenwich.

These decisions, taken in the form of resolutions with varying majorities, did not come out of the blue. Fierce discussions took place at the conference, not about the desirability for a standard time as such, but about the proper location of the prime meridian. At an early stage in the debate, the majority of those present were already in favour of Greenwich; the representatives from France, however, found it difficult to join with this position. They argued that, first of all, it was not yet the appropriate moment to discuss the location of the meridian because the principles of the system of standard time had not been sufficiently addressed. Secondly, they found that if a decision must be made, then a 'neutral' location should be given preference to the capital city of the already powerful England. Once the French realised that they were not going to acquire a majority in favour of their position, however, they supported a suggestion from Spain: in return for a prime meridian running through Greenwich, the British would agree to change over to the metric system. In response, the British spokesman stated that no matter how much he might welcome the idea, he could not promise that he would be able to persuade the people of Britain to adopt it. In the end, twenty-two votes were cast in favour of Greenwich; France and Brazil refrained from voting; San Domingo voted against the resolution.[20]

England, therefore, gloriously carried the day over France, but it could not attribute its victory to the power of the British Empire alone. Rather, the decisive factor in the arguments was based on international navigation and the fact that for over a century sailors had been using Greenwich time as the standard time— a standard of particular relevance for determining the east-west position of

20 Howse 1980: 138-151.

ships. As early as 1880, two-thirds of all merchant sailors in the world were already orienting themselves according to Greenwich time.

The dispute took place exclusively between the European powers of England and France. The United States did little to draw attention to themselves and solidly took the side of England. This does not mean, however, that they played a minor role. Not only were they the initiator and host to the conference, but they had already set an example in the previous year by introducing a domestic system of time zones based on the Greenwich meridian.[21]

In retrospect, this was the pivotal moment: the decision made in 1883 by the major railway companies in the United States to divide the country into four time zones. This set a precedent which apparently worked. The precedent as such, of course, was firmly grounded in its own history: firstly, the emergence of a national network of railways with all the co-ordination problems it entailed; secondly, the long-established practice found in all localities along every railroad of measuring local time in standardised hours and minutes; and thirdly, the availability of the system of degrees of longitude used by sea navigators with Greenwich as the point of orientation.

The railways, not the government, implemented a single national standard time in the United States. It was no different in other countries. The railways took the lead and the national legislatures followed. In the Netherlands in 1858, for instance, the director of the Rhine Railway declared Amsterdam time as the national railway time, and this was gradually accepted by other cities as the official time. In 1892, the directors of the railways—under pressure from their German colleagues—changed over to Greenwich time, and the post and telegraph followed. Not until 1909 was a single standard time established by law for the whole country; this, however, was not Greenwich time, but Amsterdam time, which ran ahead of Greenwich time by nineteen minutes and thirty-two seconds. To make international co-ordination a little easier, on July 1, 1937, the time difference was set at an even twenty minutes. On May 16, 1940, the German occupying forces imposed Middle European time on the Netherlands, making it the last country in Western Europe to join the system of time differentials based on the full hour.[22]

Each land in Western Europe developed toward the standard system according to its own course and pace. Nearly everywhere, voices of resistance were raised to condemn the unconditional approval of 'English' Greenwich time.[23] At the same time, however, people in the second half of the nineteenth century were well accustomed to the pattern of standardised hours and minutes, as it was

---

21 Howse 1980: 124.
22 Knippenberg and De Pater 1988: 80-82.
23 Cf. Kern 1983: 13-14.

already widely used at a local level. Because of this, the step from local, to national, and then to international standard time did not turn out to be very difficult.

Nowadays, the idea that one is submitting to British hegemony by accepting the Greenwich meridian as the prime meridian rarely comes up, at least not in Europe. Outside of Europe, occasional protests continue to be heard. In 1979, the Ayatollah Khomeini called it an unacceptable thought, even a nightmare, that he had to be subjected to the clock time of Europe.[24]

Indeed, the division of the world into time zones is oriented around Europe. It is probable, though, that the twenty-four-hour system originated in the Middle East—out of Khomeini's neighbouring country, Iraq. Within Europe, the final word on standard time is no longer determined in Greenwich; since 1913, that has been taking place in Paris, in the Bureau International de l'Heure. The French institute does not play a significant role in the daily practice of timing around the world, however. The time grid functions autonomously, without regular guidance from Greenwich, Paris, or any other centre.[25] The Eurocentric character is most evident in regard to the date line: this is the invisible line which runs along the one hundred and eightieth longitude, as far away from London as possible—and, because of which, the Fiji Islands are always one day behind neighbouring Samoa.

The Contemporary Time Regime

It is conceivable that at some point in time, a tyrant of a military-agrarian empire could have attempted to set the time according to his own will, and all his subjects would have been forced to conform the division of their days to match his. It still happens that people make other people wait—sometimes for hours. The length of the hour, as well as the beginning of the hour, though, are indisputable. More than anything nowadays, time itself is the tyrant; nobody anymore has the illusion that they can set it their own way.

Uniform clock time is linked to some longer-term divisions of time. One of these, the combination of the Christian era and the Gregorian calendar, has become the standard for most parts of the world. The fit between the days, weeks, months, and years is irreparably awkward; still, wherever the system is used, it provides people with a convenient way of reaching agreement about the time of day, date, and year.

---

24 Zerubavel 1982: 19.
25 Officially, Greenwich Mean Time (GMT) has been replaced by the Co-ordinated Universal Time as the international standard time, determined in the Bureau International de l'Heure in Paris, and indicated by the initials UTC. The differences among GMT and UTC are very small. The same goes for the third world standard time, International Atomic Time (TAI), which is also determined in Paris. For further details, see Howse 1980: 173-190, and Macey 1994: 156-159.

In one way or another, the naming of the globally accepted system of time continues to reflect its origins. The names of days and months, in many languages, are directly reminiscent of Roman gods and emperors, and the most prominent calendar is 'Christian.' Attempts have been made to obscure the Christian origins by replacing the words 'before Christ' with the initials BCE (Before Common Era). This substitute, however, does not do very much in removing the original orientation toward the Christian tradition. The same goes for the prevailing geological time scale. Major eras continue to be named after the European areas and regions where their traces were first discovered and described: Jura, Devon, Maastricht, and so forth.

The development of many new techniques has made it possible to spread the uniform time grid further and to expand it extremely far into the past of historical, archaeological, geological, and astronomical time. Historians can go back five thousand years, to the earliest written sources; archaeologists, several hundreds of thousands of years to the oldest remains of human culture; geologists, more than five billion years to the creation of the earth; astronomers, somewhere along the lines of fifteen billion years to the Big Bang. Any type of past that we can possibly conceive of is caught up into the same net of time, and accessible to 'absolute dating.' All processes which have ever taken place can be expressed in terms of the same convertible units of time.

At the same time, this time grid can be broken down into the smallest units imaginable. The standard for contemporary chronometers is set by the cesium atom clock, measuring to the 9,192,631,770 fraction of a second.[26] The accuracy of such clocks is so great that they can determine the most minute deviations in the rotation of the earth on its axis. The clocks themselves are said to have a deviation of no more than one second in 350,000 years.

The time grid in its entirety is uniform and anonymous, and applicable to the past, the present, and the future. Any and every moment can be assigned to its place in the system of years, days, hours, minutes, and seconds. In offices and factories, in observatories and laboratories, at the stock exchange and in markets, in international treaties, and for athletic competitions—people unquestioningly avail themselves of one and the same time scale. All measured times are internationally comparable and all are subjected to the ever-increasing precision of chronometers; without this, no athletic world records would be possible. Moreover, the time grid is also strictly synchronised so that it is possible to determine exactly whether a transaction in Tokyo took place a second earlier or later than a similar transaction in New York.

Timing at the end of the twentieth century has achieved an unprecedented degree of precision, orderliness, range, and relevance.[27] The degree of precision

---

26 Macey 1994: 158. For a clear explanation, see Barrow 1991: 130-131.
27 I have used these four concepts as ordering principles in Goudsblom 1977.

extends to the millionth of a second. Orderliness, however, continues to be hampered by the fact that a year is not a simple multiple of days—a fact which obstructs a neatly metric or duodecimal system of nicely rounded-off numbers. Here, we can clearly see that the human intellect is not sovereign in marking time. On the one hand, the movements of the earth allow us to construct a very simple arithmetic relation between years and centuries. On the other hand, they make an equally simple division of the year into days and hours impossible. In this respect, nature sets limits to culture. Culture itself, moreover, exerts a robustness of its own in the survival of the Babylonian duodecimal system.

Even if a few knots in the net of time have thus far resisted a simple arithmetic solution, the range of the net has, in spite of these imperfections, expanded over the whole earth with seemingly no problems. It has at once become more intricate and more comprehensive. It extends from the furthest point imaginable in the past to the furthest point imaginable in the future. It also includes, in principle, mental processes—even though these may be experienced in a very different way.[28]

The compelling strength of the global time grid lies in its relevance—in the functions it has for all those who are involved in it and collectively maintain it. As its last step, the time grid forms a time regime which exists by virtue of the continuing pressures that people exert on each other and on themselves. People arrange their lives according to the clock because they know that others do also. In this way, as early as the seventeenth century in Europe, the penetration of the clock had taken on its own dynamic; in the words of the Italian historian Carlo Cipolla:

> ... the machine which had been devised to satisfy particular human needs created new ones. Men began timing activities that, in the absence of clocks, they had never thought of timing. People became very conscious of time, and, in the long run, punctuality became at the very same time a need, a virtue, and an obsession. Thus a vicious circle was set into motion. As more and more people obtained clocks and watches, it became necessary for other people to possess similar contrivances, and the machine created the conditions for its own proliferation.[29]

Cipolla takes a negative view of the development. His statement expresses a widespread tendency to experience the time regime of clocks as a fatal fetishism: people have turned the clock into an idol which they are now forced to worship.[30]

---

28 Adam (1990: 129) cites the description of a person's experience in the face of death. It was said that the experience 'seemed to last for centuries,' but in reality it took only 'a few seconds.'
29 Cipolla 1967: 103.
30 Cf. Young 1988: 227-228.

It was in a similar vein that Evans-Pritchard considered the Nuer People fortunate for not knowing time.[31]

Time manifests itself as a tyrant wherever people compel themselves and others to obey it. The constraints they experience as time pressure are the workings of the time regime—the urgency they feel to be somewhere or to be done with something 'on time.' As Norbert Elias noted in his study of the civilising process, the pace at which people live is a function of the amount of social connections in which they are involved.[32]

This pace is now picking up impulses from all over the world. A century ago, the railway stations functioned as the principal nerve centres of the new national times which, in turn, quickly assumed an international dimension. Nowadays, there are numerous international organisations with national bases at airports, in embassies, at stock exchanges, and in offices, which together form just as many junctions in the network of global time. The network finds further support in a multitude of direct personal connections by way of radio and television, the telephone, and computers. Satellite and cable connections reduce the time needed to make contact to a minimum.

The result is an abundance of information circulating around the world at unprecedented speed. Almost inevitably, this tempo leads to a shortening of time perspectives: one has to be constantly informed about the latest developments and events, as well as be prepared to react to them promptly. The need to keep abreast is unrelenting.

The extent and the intensity of the interdependencies, at the same time, also foster a lengthening of time perspectives. The same combination of knowledge and skills which allows people to reach further and further into the past, also enables them to extend the time grid into the future. In economic forecasting, scenarios are usually made in terms of months or years; demographic predictions can be made in terms of generations; and ecological projections are, at times, already couched in terms of centuries. Experts active in computing the radioactivity of nuclear waste do not even shy away from estimates along the lines of half a million years.

These forecasters and experts all operate within a linear concept of time. Of course, the years which are their standard units refer to regularly repeating processes of a cyclical nature. They balance, however, those years against a linear time axis according to which no single moment can ever repeat itself.

The art of timing continues to rest on the principle of repetition. All the technical devices which people have developed for measuring time either represent or reproduce regularly repeating processes, the speed and duration of which are known. Aided by these devices, physicists have designed an image of the uni-

---

31 Evans-Pritchard 1940: 103.
32 Elias 1994: 457-458.

verse as a clock, subjected to eternally unchanging laws of motion. Over the last two centuries, an awareness of the limitations of this image has grown, first in geology and biology, and later in astronomy and physics. In all these sciences, a 'rediscovery of time' is taking place, resulting in a view of the entire universe within the framework of an all-encompassing singular evolution.[33] It is, perhaps, not far-fetched to suppose that a link exists between this 'temporisation' of the world image and the globalisation of the time regime.

33 Prigogine and Stengers 1984: 26.

2 Modern Times in Bangladesh

Willem van Schendel

Time can be studied in many ways. In this chapter, I treat it as a social category which expresses the human experience of duration and irreversibility, and as a social construct which varies from one group of human beings to the next. It rests 'upon rather arbitrary social conventions,' which are themselves subject to change.[1] In this way, the study of constructions of time may contribute to a better understanding of social change.

There are three strong, but largely separate, trends in the literature on time as a social construct: explorations of the 'global time regime' (mainly by sociologists), studies of particular time regimes in different cultures (mainly by anthropologists and social psychologists), and studies of periodisation in professional history writing (mainly by historians). In this chapter, I examine the possibility of combining these trends in order to paint a picture of how constructions of time have an impact upon each other in particular historical settings. My canvas explores one particular society, Bangladesh, during the second half of the twentieth century.

To begin, it is important to get away from two conventions in the literature. Firstly, studies of the global time regime remain strongly Eurocentric and convergence-oriented. They are eager to highlight global unity but are less impressed by, and informed about, local difference, which is hastily seen as archaic, non-progressive, pre-modern, and on the way out.[2] Recent studies (e.g. Young, Zerubavel, and Goudsblom) have explored the emergence and consoli-

---

1 Zerubavel 1981: xii.
2 Adjaye speaks of 'either outright denial or distortion by Western authors' on the subject of time among Black peoples in Africa and the diaspora. Adjaye 1994.

dation of a 'global time regime' as the outcome of a complex of social, economic, and technological changes in industrialising societies.[3] This complex gave rise to a 'metronomic society' ruled by clocks, calendars, fixed working hours, and timetables, which then spread to envelop the entire human race. A prominent symbol of this global time grid is the Greenwich-oriented international system of time zones. Academic historiography and the literature on globalisation and 'modernity' have also strongly fostered our sense of global or 'world time.'[4] But these studies provide little information as to how a world time regime was imposed and contested, and whether it has in fact been as successful as they claim in organising the life of human groups world-wide. How did the world time regime mesh with numerous local time regimes all over the world? What were the conditions in non-industrialised societies that allowed global time to be integrated into, or appropriated by, local times? How did local times shape the present form of the global time regime? Which forms of resistance to world time could (and can) be identified?

Secondly, the literature on time regimes in different cultures tends toward localism and particularism; it treats local time regimes as relatively closed, integrated, and static. The interactions between local regimes and the unfolding world regime remain underexposed, including the possibility that localisms may be recent results of these interactions rather than age-old cultural elements.[5]

Additionally, it is important to examine the middle ground between the assumptions of linear convergence and particularism. How have constructions of time been handled by social actors in particular historical situations? How and why have these actors manipulated the time regimes at their disposal, leading to new forms, mergers, and hybridisation? And which modern times actually exist? This approach is likely to show that the hegemony of a 'global time regime' is much less noticeable on the ground than one might deduce from the literature on globalised time, and that the dynamics of change do not necessarily point in the direction of convergence.

In this chapter, I do not start from the evolutionist, globalist, nor convergence-oriented view, nor do I posit any closed system of 'local time.' I look, rather, at the existence of multiple constructions of time in Bangladesh, not as 'pre-modern' survivals but as an expression and outcome of cultural and political struggles. Time is not just imposed by a globalised world elite (or by the faceless structures and institutions of which they are the visible representatives); time can also be a tool in the hands of local groups seeking to impose their own agency on history.

---

3 E.g., Zerubavel 1981; Young 1988; Goudsblom 1997.
4 On the concept of world time, see Skocpol 1979. On temporality and modernity, see contributions to Featherstone, Lash, and Robertson 1995.
5 Hoskins 1993.

This chapter sketches three arenas in which time became a contested entity in twentieth-century Bangladesh. In all three cases, the struggle over time was linked to changes in the position of elite groups in society. These arenas are *identity politics, historiography,* and *development discourse.* But first let me say a few words about the complexity of the local time regime.

A Time-rich Society

People in Bangladesh live in a time-rich social environment: they have long had at their disposal several ways of ordering time. As in many other Asian societies, various measurements of time co-exist. For example, in a widely used almanac for the year 1998-1999, no less than *fourteen* different calendrical systems are mentioned.[6] Clearly, not all of these are in regular or general use. Most people in Bangladesh, however, routinely use three different ways of measuring time. These are known as Bengali Time, English Time, and Islamic Time. The first two calendars are solar, and fixed to one another.[7] The Islamic calendar is lunar, and shifts with regard to the other two.

The Bengali calendar has been used to structure time in Bengal for many centuries. It is a deeply meaningful cultural artefact. Cultivators attune their actions to its six seasons, and agricultural wisdom is contained in many expressions involving the Bengali calendar. The state collects land tax according to the Bengali months.[8] For most Bangladeshis, it is of great importance to choose auspicious days and months for the celebration of life-cycle events, such as marriage or the feeding of the first rice, according to the Bengali calendar. Great events and calamities are popularly remembered in Bengali Time—the devastating Great Famine of 1350 occurred in AD 1943. Books published in the Bengali language are usually dated according to Bengali Time (BE = Bengali Era), and the rich poetry of Bengal abounds with references to Bengali Time. The national anthem of Bangladesh, for example, composed by Rabindranath Tagore, begins with the following lines (names of months not in italics):

---

6 The Loknath Almanac opens with the statement that this is the almanac for the Bengali year 1405, which corresponds to AD 1998-99, 1418-19 Hijri, 1920-21 Saka, 2055-56 Sangbat, 1405-06 Fasali and Amali, 13560-61 Magi, 1406 Bagri, 1408 Tripuri, 513-14 Chaitanya, 268-69 Loknath, 187-88 Hari, 549-50 Kamrupi Sankara, and 2541-42 Buddhist era (*Loknath* 1998, 1).
7 The Bengali year starts in the middle of the month of April of the English year (= Common Era [CE] or AD); all twelve Bengali months run from about the middle of an English month to the next. The Bengali calendar is 593 years behind the Common Era: April 2001 corresponds with the beginning of 1408 BE (Bengali Era).
8 On the agricultural calendar (*baromashi*), see Zbavitel 1963.

My golden Bengal, I love you.
Your skies, your breezes, ever with my breath play the flute.
O mother, in Phalgun *the perfume of your mango groves drives me mad.*
Ah mother, what honeyed smile have I seen on your laden fields in
Agrahayan.[9]

In a similar vein, a well-known contemporary poet, Taslima Nasreen, marked the Bengali year 1400 (AD 1993-1994) with a poem entitled '1500,' in which she imagined Bangladeshi society a hundred years from now.[10]

Time and Identity Politics

Bangladesh differs from other Asian societies in the complexity of its de-colonisation. Having been released from British rule in 1947, it found itself a dominated society under Pakistani rule. Bangladeshi intellectuals usually describe British and Pakistani imperialism as similar systems of racialised domination.[11] It was only in 1971 that Bangladesh came to be an independent state. This unusual trajectory has produced a fractured history of identity politics, which differs markedly from that of other parts of erstwhile British India.

In the process, Bengali Time became *one of the symbols of Bengali nationhood*. It is significant that the Common Era is still known in Bangladesh as 'English Time.' Introduced by a colonial state elite, it retained strong foreign and Christian connotations and never became fully naturalised. The use of Bengali Time played a symbolic role in the anti-colonial struggle against the British (the national anthem quoted above was composed in the early 1900s), but it became much more important during the period in which Bangladesh (then East Pakistan) was part of Pakistan (1947-1971).

During this period, a number of symbols were used to evoke the regional identity of the people of East Pakistan. The most encompassing was the image of Golden Bengal (Shonar Bangla), a cultural construct of prosperity, indulgence, abundance, and beauty. Often anthropomorphised as 'Mother Bengal,' this essentially timeless image had been able to fire intense emotions among Bengalis in the struggle against British rule in the early twentieth century, and it proved to be equally powerful in mobilising Bengalis against Pakistani rule in the 1950s and 1960s.[12] The leader of the movement for Bangladesh, Sheikh

---

9 Translation by Edward C. Dimock, Jr., and Ronald B. Inden; see Greenough 1982: 9.
10 For an English translation, see Nasreen 1997: 94-96.
11 E.g. Jahangir 1998.
12 When Bangladesh gained independence in 1971, it adopted the national anthem quoted above; it is entitled 'Amar Shonar Bangla' (My Golden Bengal). For an interpretation of the cultural meaning of Golden Bengal, see Greenough 1982: 12-41.

Mujibur Rahman, did more than invoke a shared conception of prosperity, a celebration of the Motherland. He called for the revival of Golden Bengal, thereby imposing a simple but effective nationalist periodisation.[13] He provided Bangladesh, then merely a 'nation-of-intent,' with a pristine and immemorial past, as well as a limitless future. When the original Golden Bengal might have existed remains vague; one surmises that it was somewhere in the pre-colonial period, but the periodisation of history was clear: 1) Golden Bengal; 2) Golden Bengal destroyed (by foreign rule and capitalist exploitation); 3) Golden Bengal restored (by the future creation of an independent, socialist, democratic Bangladesh).

The use of Bengali Time came to be seen as one mechanism for restoring Golden Bengal. Using it was a political act, pointing as much to a glorious past as to a glorious future. Celebrations of important days on the Bengali calendar (such as Pohela Boishakh, or Bengali New Year [April 14]) were organised as manifestations of cultural pride and resistance against 'Pakistanisation.'[14] The emotive potential of Bengali Time was fully exploited in the identity politics of the 1950s and 1960s, and the state of Bangladesh, which emerged in 1971, decided to use Bengali Time in official documents and circulars. Today, it continues to be widely used, although, with the receding of Golden Bengal as an image of hope and the emergence of new symbols of identity, it has become less prominent in the fields of politics, administration, business, and development activities.[15]

Bengali Time was used effectively in the bid for Bangladeshi nationhood in another way as well. It acted as an 'homogeniser' of time in several regions of Bangladesh where other calendars were in use. As such, it was part of the larger cultural offensive to naturalise control from Dhaka over the territory. Large parts of the eastern districts of Comilla and Noakhali, for example, had been administered by neighbouring Tripura state, and used Tripura Time.[16] Several non-

---

13 E.g. in his speech 'Shonar Bangla Shoshon Keno?' ('Why is Golden Bengal a Graveyard?'); see Jahangir 1998: 4.
14 Bengali New Year became an elite celebration. For most other Bengalis, especially in the rural areas, it had long had other connotations as the day on which merchants and shopkeepers would call in the previous year's debts and open a fresh accounting book for the new year (hal khata, or 'new ledger').
15 In the media, three calendars are widely used. Each day, television transmission begins with an announcement of the date according to Bengali, Islamic, and English time, and newspapers often carry several parallel dates. The daily Internet edition of the English-language *News From Bangladesh* opens with the Bengali date, followed by the Islamic and the Common Era date (www.bangladeshweb.com/news/). Other Internet editions list Bengali and Common Era dates (e.g. the English-language *The Independent* [independent-bangladesh.com] and the Bengali-language *Muktakantha* [www.muktakantha.com]), thereby signalling to their readers that their politics are not fundamentalist.
16 E.g. the Chakla Roshnabad Estate, the extensive *zamindari* domains of the Maharaja of Tripura in British Bengal.

Bengali groups in the territory of Bangladesh have their own systems of calculating time. The Khasi of north-east Bangladesh use a lunar calendar with an eight-day week.[17] The Marma of the far south employ their own calendar which is closely related to the Burmese calendar.[18] Such regional calendars have been under attack, but they persist as long as they are linked to important economic activities (as in the eight-day sequence of local day-markets among the Khasi) or socio-religious events (as in the Buddhist rituals of the Marma).

In the post-1971 period, Bengali Time took on a new symbolic role in the identity politics of the independent nation. Now, Bengali Time turned into a *symbol of secularism*. A major issue in national debates about Bangladeshi identity was the extent to which either regional culture or Islam should be emphasised. As there is both a Bengali calendar and an Islamic one, the strategic use of one or the other could signal different positions in this debate.

An example of the subtleties involved in this is reflected in the position taken by successive Bangladeshi governments. Between 1975 and 1996, Bangladesh was ruled by governments which, although not actively Islamist, made frequent overtures to Islamist forces in the country. They emphasised the fact that most citizens of Bangladesh are Bengali-speaking Muslims, and they used this to distinguish the 130 million Bengalis in Bangladesh ('Bangladeshis') from their 100 million fellow Bengalis in India ('Bengalis'). Secularists in Bangladesh objected to this distinction, which they considered false and harmful; they foregrounded the unity of all Bengalis as a cultural community. As a result, the distinction between 'Bangladeshi' and 'Bengali' became a sticky political issue, and governments made various efforts to highlight 'Bangladeshiness.'

Some of these efforts had to do with the manipulation of time. The Islamic weekend, for instance, which focused on the Friday, was introduced to replace the Christian/English weekend, which focused on the Sunday. As the majority of inhabitants of Bangladesh are Muslims, and the rhythms of the Islamic annual cycle dominate public life, few people object to foregrounding Islamic Time. But there are sizeable religious minorities in Bangladesh, notably Hindus, Buddhists, and Christians. Indeed, their calculations of time exert an influence on public life as well, if only because the state has declared various non-Islamic religious events (such Christmas, or Buddho Purnima) as public holidays.[19]

Apart from the use of Islamic Time next to Bengali Time, the government also sought to use Bengali Time in its quest for a Bangladeshi national identity. In the mid-1990s, it introduced a specifically Bangladeshi Bengali Time, which differed from the Bengali Time used in neighbouring West Bengal and Tripura

---

17 Lyngdoh 1986.
18 Bernot 1967: 167-197.
19 Such public holidays were often demanded in petitions from minority groups to the government. See e.g. the proceedings of the Buddhist Conference, Chittagong (National Archives of Bangladesh; Government of East Bengal, Department Home [Political] B. Proceedings, 9B-8 of 1951).

(India) by one day.[20] The idea was to bring the calendar more in line with the solar year, and to do away with awkward features such as a 32-day month. One result, however, was that it created great problems for the millions of Hindus in Bangladesh, for whom the Bengali calendar is also a ritual calendar linked to astrological and cosmological observations.[21] Rather than take the ritual risks involved in switching to the new Bengali Time, the vast majority of Bangladeshi Hindus decided to stick to the old calendar for the timing of their religious activities. As a result, there are now two Bengali calendars in use in Bangladesh, one for general purposes and another—with a one-day shift—for ritual purposes.

Bengali Time has also been used by political parties to sway public opinion in their favour. For example, the Awami League supports the celebration of 25 Boishakh as the birthday of Rabindranath Tagore; the Bangladesh National Party supports the celebration of 11 Joistho as the birthday of Nazrul Islam (another famous poet who was declared National Poet).[22] Since 1996, when the Awami League returned to power after 21 years in the opposition, Bengali time finds itself once again backed by the state. In 1999, on the first day of Spring (February 14), a National Committee for the Spring Festival organised a day-long programme including dances, processions, a fair, poetry readings, a musical soiree, and an exchange of flowers (Figure 1).[23] Bengali New Year (April 14) was celebrated as a powerful symbol of secularism and nationalism. A procession carrying 'a replica of a big black snake, symbolising fundamentalism and communalism, and white pigeons symbolising peace' was brought out in Dhaka.[24] The festivities were given live coverage on national television, and newspapers published special articles on the occasion. Some of these pointed explicitly to the importance of the Bengali calendar in the cultural defence against globalisation.[25] Figure 2 shows the New Year's procession in Dhaka.

20 The official homepage of the state of West Bengal features a detailed Bengali calendar for the current year (see http://host.westbengal.com/misc/calendar).
21 For an analysis of Bengali Time in Hindu ritual terms, see Östör 1980: 212-214 ('The System of Time').
22 Rabindranath Tagore was born on May 7, 1861, and Nazrul Islam on May 24, 1899.
23 'Pahela Falgoon Observed,' The New Nation (Dhaka), February 14, 1999; 'Basanta Utsab at JU,' Daily Star (Dhaka), February 16, 1999.
24 'Pahela Baishakh Today,' The Independent (Dhaka), April 14, 1999.
25 'Pahela Baishakh is the first day of the Bengali calendar year. As is the case with the first day in the calendar of other peoples (meaning calendars other than English), Pahela Baishakh is a day of special significance. The day represents all that the Bengali culture stands for [...]. [T]he day has a symbolic nation-building relevance [...]. For a Third World aid-dependent country the degree of political autonomy is always open to question. But what about cultural autonomy? [...] Can we sustain our cultural heritage in these days of cultural globalisation? [...] [H]ow far the observance of Pahela Baishakh would mean a demonstration of our cultural autonomy?' (Syed Anwar Husain, 'Pahela Baishakh and the Challenge of Globalisation,' The Independent [Dhaka], April 14, 1999). Another article pointed to the need 'to introduce the city dwellers with traditional Bengali heritage.'

Figure 1
Girls dressing up for the Spring Festival. Dhaka, Bangladesh, 1 Phalgun 1405 B.E. (= 14 February 1999).[26]

As a result of the multi-temporality of identity politics in Bangladesh, the official calendar has become highly complicated. This is not only because there are special days to celebrate state occasions, cultural events, and the sacred days of several major religions, but also because these are expressed in dates derived from different calendars. Thus, cultural celebrations such as the great poets' birthdays or the beginnings of the seasons are expressed in Bengali Time; Islamic festivals (e.g. Id-ul-Azha), in Islamic Time; and state events (e.g. Victory Day [December 16]), in English Time.

Clearly, in Bangladesh, the measurement of time has become linked with political struggles. Bengali Time became a symbol for the aspirations of the Bengali regional elite in Pakistan. This elite rejected the fact that the state of Pakistan was dominated by a West Pakistani elite and their symbols. It used cultural symbols to galvanise Bengalis into supporting its bid for power. Much attention has been paid to the use of language in this struggle—the Bengali autonomy movement is often referred to as the Language Movement—but other sym-

('Pahela Baishakh Celebrated: Week-long Fair Opens,' *The Independent* [Dhaka], April 16, 1999). Cf. 'Bangla New Year: Pledge To Uphold Bengali New Year,' *The New Nation* (Dhaka), April 16, 1999.
26 Source: *The New Nation* (Dhaka), 14 February 1999.

MODERN TIMES IN BANGLADESH

Figure 2
Procession celebrating Bengali New Year. Dhaka, Bangladesh (photo: Shafiqul Alam Kiron/Map Photo, Dhaka).[27]

bols were also employed. Among these were dress, food, and time. As indicated above, these symbols were used to fight the Pakistani state elite, but also to consolidate and legitimate the Bengali regional elite's power over others in what would become Bangladesh.[28]

In the process, Bengali Time came to be associated with religious tolerance, liberal notions of Islam, and progressive politics. When Bangladesh came into existence, Bengali Time experienced a period of prominence. But with the waning of the liberal trend in Bangladesh beginning in the mid-1970s, its temporal symbol also found itself under attack. A main contender was Islamic Time, put forward by Islamist and conservative politicians as the more appropriate measurement of time for Bangladeshis. The celebration of time-specific Islamic ritu-

---

27 Source: *The New Nation* (Dhaka), 16 April 2000.
28 A danger of linking the measurement of time with a political struggle is, of course, that a calendrical system can be completely abandoned if its political cause is defeated. This is what happened to the republican calendar created by French revolutionaries, which introduced a new way of measuring time, thought to be at once more rational and emancipatory, and which started from the year of the revolution (A.D. 1789 = The Year I). On its abolition and disappearance from French popular memory, see Baczko 1997.

als (e.g. daily prayers, Friday prayer, Ramadan, etc.) became one yardstick against which people's politics could be measured.

In this major conflict, the global time regime ('English Time') was of little account. Disliked by most as a Western and Christian import, and admired by others as a symbol of modernity and global 'cool,' it was used as a politically neutral system.[29] Despite an increased use of wrist watches and clocks, and the existence of printed calendars and transport timetables showing global time, the significance of global time for most Bangladeshis should not be exaggerated. Global time remains one among several options, and most Bangladeshis are multi-temporal.[30] Through the Bengali month and the lunar religious year, large areas of life are measured by the *bela* (a four- to six-hour part of the day), rather than the hour or the minute. Moreover, merely using global time obviously does not imply that a society has become a 'metronomic society.' Clocks are often treated more as ornaments than as stern task-masters. Overt signs of 'metronomy' may hide a very different reality. What 'time literacy,' to use Robert Levine's term, requires is that we distinguish between clock time (in which the hour on the timepiece governs the beginning and ending of activities) and event time (in which scheduling is determined by activities).[31] In Bangladesh, as elsewhere, concepts of global time are domesticated and adapted to local needs. Local and global concepts of time are continually manipulated and reinvented to serve new political needs, and to give expression to social identities. As long as multi-temporality serves such purposes, it will flourish.[32]

State Time: Constructions of Bangladeshi History

The second arena in which time became contested in Bangladesh was history writing. In what to many Bangladeshis has been a long period of disorder and dashed hopes, history was used as a tool for coping with shared experiences, val-

---

29 The global time grid—the system of time zones with a standard variation from Greenwich time (Universal Time Coordinated [UTC])—seems wobbly in Southern Asia. Countries like Iran, India, Nepal, Sri Lanka, Burma, and Indonesia deviate from the grid. In this respect, Bangladesh appears as an island of rectitude: since 1951 it follows the international system and as a result is 30 minutes ahead of India and 15 minutes ahead of Nepal. In October 1951, the Government of Pakistan introduced the East Bengal Standard Time, being six hours in advance of Greenwich Mean Time. Earlier, clocks in East Bengal (now Bangladesh) had been running five-and-a-half hours in advance of Greenwich Time (National Archives of Bangladesh; Government of East Bengal, Department Home [Political], B. Proceedings, 16A-7 of 1951).
30 Levine 1997: 217-220.
31 Levine 1997: 85.
32 A parallel is the continued use of other regional measures in Bangladesh, e.g. *lakh* (100,000 units) and *crore* (100 *lakh*).

idating current behaviour, and guiding future action. Little has been written, however, about the concepts of time that have informed the study of history.

It is not possible simply to record the passage of time; we pattern our understanding of the past on how we experience the present. The study of history involves selecting events, constructing chronologies and processes, mapping out periods, and imputing causes and effects. Without such manipulations, the past would remain meaningless, chaotic, overwhelming, and inaccessible. If we are to 'make sense' of the past, bring meaning and order to it, we must structure it into coherent sequential accounts. But we must also reflect upon, and continually reassess, the sequences offered to us. One way of doing this is by focusing on the cultural, generational, and individual aspects of the time patterns that can be found in the work of historians. This can take the shape of examining how historians have presented 'processes,' 'events,' or 'causality' as tools for knowing the past. Here, however, we shall look briefly only at periodisation.

Historians of Bangladesh share a strong commitment to a linear understanding of time. They write about 'the time we have overcome', a past which lies 'behind' us. While this is the mode of almost all contemporary academic history writing, there are differences in emphasis. For example, Burmese historiography has been somewhat more inclined to impute repetition and cyclical or spiral understandings of historical processes than Bangladeshi historiography.[33] In Indian historiography, the question has been raised as to what extent the linear concept of time has been part of a 'bourgeois narratology,' and whether 'chronology itself, the sacred cow of historiography, [...] [should] be sacrificed at the altar of a capricious, quasi-Puranic time which is not ashamed of its cyclicity.'[34] Such cyclical notions are absent in the historiography of Bangladesh. Remarkably, unlike politicians and artists, historians have never bothered to develop the cyclical interpretation of history offered by Sheikh Mujibur Rahman.[35] If anything, they have ignored the notion of an historic Golden Bengal projected onto the pre-colonial past.

Bangladeshi historiography is not just linear in the sense that historians conceive of time as irreversible, but also in that, to most of them, it has shown a clear direction and is, by implication, a relatively transparent guide for current action. Presented in this way, history provides a satisfying clarity that the present has often been lacking. This directionality is shared by sequential accounts which highlight such different themes as the penetration of capitalism, the progress of modernity, or the emancipation of the Bengali/Bangladeshi nation.

In order to make time manageable, historians have always found it helpful to break up their sequential accounts into time segments, or periods. A period—

---

33 E.g. Lieberman 1984.
34 Guha 1996: 12.
35 The 'folk-based secularism' of artists like Zainul Abedin and Jasimuddin strongly supported the vision of Mujib. See Jahangir 1993; Jahangir 1998.

assumed to bracket important continuities and processes, and to be separated from earlier and later periods by significant discontinuities or disruptions—is not preordained, but a matter of considered choice: it is most effective if related to the problematic being explored. Historians of Bangladesh have shown a striking unanimity in the way they have constructed their periods.

If we liken the historiographical landscape of Bangladesh to a mountain range stretching out into the far distance, we can discern four giant peaks towering over all others. These can be coded in different ways. In years of the Common Era, they are 1204, 1757, 1947, and 1971. In terms of disruptive events, they are the conquest of the Sena capital by Turkish cavalry, the battle of Plassey, the Partition of British India, and the Liberation War. In terms of emblematic historical figures, they are Muhammad Bakhtiyar, Nawab Sirajuddaula, Mohammad Ali Jinnah, and Sheikh Mujibur Rahman. The mountaineers of Bangladeshi history have used these codes for the giant peaks to demarcate the territories lying in between. In turn, they have developed these territories into fields of specialisation, largely confining themselves to exploring the lower peaks lying in between the giants. In this way, the last 800 years of Bangladeshi history have been parcelled out as Muslim or pre-colonial (1204-1757),[36] British or colonial (1757-1947), Pakistani (1947-1971), and Bangladeshi (1971 to the present).[37]

These periods have become so familiar that few historians feel a need to justify or defend them, and studies embracing more than a single period are extremely rare. The assumption is that scholar and audience share an understanding of the significance of the disruptions which mark the cut-off points of each period. However, the periodisation of Bangladeshi history is predicated upon political disruptions and a statist perspective, and may not be all that useful for historical problematics which do not privilege state politics. This point was raised by Richard Eaton in his groundbreaking study of the rise of Islam in Bengal between 1204 and 1760. He stated that 'one may legitimately ask why a cultural study ends with a political date,' and provided two reasons for why this can be done. His first reason is that the political disruption at the end of his period of study, the coming of the English East India Company to rule Bengal, also

---

36 For an influential account, see Sarkar (ed.) 1948.
37 For example, Shireen Hasan Osmany distinguishes five phases of Bengali Muslims' awareness of 'their distinct and different identity': 'The first phase lasted from about the eighth century to 1204: i.e. Muslim conquest of Bengal. The second phase, 1204-1757, ended with the loss of Muslim political power in Bengal to the East India Company in the battle of Plassey in 1757. The third and the fourth phases ran almost concurrently during British rule up to 1947. The third phase of Muslim consciousness in Bengal was typified by the response of the Muslims of Bengal to their loss of political power to the British. During the same period the fourth phase of Muslim consciousness evolved through the process of response of Muslims of Bengal to Hindu ascendancy in the economic and intellectual spheres of Bengali life under British patronage. The political experience of Bengali Muslims in the polity of Pakistan between 1947 and 1971 marked the fifth phase in the evolution of the consciousness of their identity and destiny' (Osmany 1992: 15).

caused a cultural one—the falling apart of the imperial 'patronage system that had played a decisive role in the articulation of both Mughal political culture and Islamic institutions.' His second reason stems from his experience of being constrained by the chronological scope of his sources: the documentation in Persian, on which he based the book's later chapters, comes to an abrupt end with the intrusion of Englishmen into Bengal's revenue system. These two reasons, one referring to a historical process coming to an end, and the other referring to a source of information about the past coming to an end, make 1760, in Eaton's reasoning, a convenient stopping point.[38]

This argument demonstrates how reflection on periodisation may raise interesting new questions about how historical processes are 'assembled' by historians, and to what extent the constraints posed by sources impose a pattern on the past which is independent of the individual researcher. Was imperial patronage really all that important for the vitality of Islam in Bengal by the eighteenth century? Eaton describes Islam towards the end of his period of study as having become 'established as profoundly and authentically Bengali,' and he points out that 'many of the social and cultural processes examined [...] continued into the later eighteenth century and even the nineteenth.'[39] How abrupt was the end of Persian documentation in view of the fact that Persian scholarship continued to get some state patronage until its abolition as a court language in Bengal in 1835? Could other sources have been marshalled to extend the chronological scope of the study? Clearly, the disruption posited by Eaton to mark the end of a period of connected time may be qualified. As disruptions are never total (if only because most people live through them into the next 'period'), it is crucial to ask what is lost and what is retained. Historical periodisation is indispensable, but always provisional.

In Bangladeshi historiography, the dominant periodisation is based on the capture of state power. This may be less than satisfactory for studies into economic, ecological, cultural, or intellectual aspects of history. It may even be constricting for the study of political processes at levels other than the state, be they local, regional, or global. Elsewhere, I have argued that regional histories in Bangladesh employ other periods. For example, historians of the Chittagong Hill Tracts in south-eastern Bangladesh have frequently used 'Jumma Time,' in which regional historical turning points are privileged.[40] Thus, a history of

---

38 Eaton 1993: xxvi.
39 Eaton 1993: 303, cf. xxvi.
40 The years 1860 and 1900 have no particular significance in Indian, Pakistani, or Bengali nationalism, but they serve as crucial markers of Jumma Time, the periodisation of history in the Chittagong Hill Tracts in south-eastern Bangladesh. Conversely, Jumma Time is not concerned with 'nationalist' turning points such as 1905 (the first partition of Bengal), 1940 (the Pakistan Resolution), or 1952 (the upsurge of the Bengali Language Movement). The ways in which time is conceptualised and ordered in regional historical narratives remains a completely uncharted area

agrarian commercialisation in Bangladesh might well stress very different crises and periods: the collapse of indigo production in the 1860s, the expansion of cash cropping in the last quarter of the nineteenth century, the slump of the 1930s, or the loss of the jute export market in the 1980s. One way of presenting an ecological history of Bangladesh could be by highlighting 'water periods' based on the effects of, for example, the Brahmaputra river seeking a new channel (the Jamuna) after 1787, rivers in north Bengal changing course after the earthquake of 1897, the completion of the Karnaphuli dam at Kaptai in 1961, or the coming into operation of the Farakka barrage in the Ganges in 1975. And a demographic history of Bangladesh could use mortality crises to separate periods: the huge famines of 1770, 1943, and 1974 are obvious candidates, but (sub)periodisation could also be based on excess mortality caused by epidemics, cyclones, population displacement, or floods.

In other words, there is ample scope for re-imagining time in Bangladeshi history and for reassessing the implications of conventional statist periodisation. The imaginary 'mountain range' of Bangladeshi history is but one of many virtual realities which we can assemble in order to give the past a presence, build coherent narratives, and create significance. The validation of each virtual reality depends on the evidence we can produce, and the range of recorded evidence is enormous. Although written documents and the narratives of earlier historians have been widely used, we have merely scratched the surface of other sources of historical information. Photographs, oral evidence, music and songs, pictures, landscape formations, patterns of vegetation and settlement, linguistic variation, dress styles, and many others remain to be explored in depth.

A moot question is why the post-modernist critique of 'grand narratives' and their linearity has hardly touched the historiography of Bangladesh. The strong emphasis on nationalist/statist periodisation may be related to several factors. First, the state of Bangladesh came into being through war in 1971, an event which is etched into the consciousness of anyone dealing with the history of the country. There is still a great sense of pain and outrage, as well as an intense urge to understand the events leading up to 1971, and the war atrocities themselves. Added to this, the memory of the war, the ambiguities of national identity, and the desired form of the new state have all been cannon fodder to fierce party politicking in the 1970s, 1980s, and 1990s. In view of the fact that most practising historians in Bangladesh have vivid memories of the entire period, and live in a political environment which continues to be destabilised by nationalist rhetoric

---

for historians of Bangladesh. Similarly, 'Jumma Space' differs from the historical geography of nationalists, not just in terms of the geographical vantage points from which history is observed, but also in terms of boundaries—between hills and plains, between ethnic groups, between rulers and ruled, between male and female space—and possibly in the very concept of space employed. In other words, historians searching for post-nationalist visions of Bangladesh need to provide us with historical analyses which allow for a plurality of time and space. Van Schendel 1998/2000.

and counter-rhetoric, it is not surprising that they have taken their cue from these debates. In other words, as children of the Liberation War, the fight over the 'grand narrative' of the nation is of such immediate importance to historians in Bangladesh that post-modernist critiques seem to offer little beyond fragmentation and confusion. It would seem that a history needs to be firmly territorialised and temporalised (i.e. nationalised) before it is ready to be disrupted. In addition, historians of Bangladesh have been much more involved in debates with developmentalists than with their colleagues in the field of South Asian history (not to mention historians working on other parts of the world). This is at least partly a result of the very limited interest that South Asianists have shown in the political and institutional circumstances under which, in South Asia, historical knowledge is produced outside the Indian academy. Such inequalities, hierarchy, and ghettoisation within South Asian studies have been detrimental to the free flow of ideas (including critiques of grand narratives, or, indeed, concepts of periodisation and time) between post-Partition academic communities.

Projected Time: Development Studies

Since the 1970s, however, the wider social relevance of historians' time has decreased considerably, having been largely marginalised by an emerging concept of time derived from the field of development studies. A number of reasons can be indicated. First, the nationalist project, with which historians of Bangladesh had so closely identified themselves, ran into serious trouble from the early 1970s onwards.[41] Second, universities and other old institutions of intellectual activity became largely paralysed for lack of funding, political infighting, periods of censorship, and a general shrinking of arenas for open debate. Third, large amounts of money became available as a result of international development aid. This inflow, largely through the state, had an effect on elite formation in the country which is perhaps not paralleled anywhere else. Bangladesh, which gained independence after an ordeal in which many of its best intellectuals were purposefully murdered, started life as an independent nation with a very small elite. Since then, the expansion of the national elite has been explosive, and almost entirely aid-driven.

Development discourse has taken over the social sciences in Bangladesh to a remarkable degree, forging strong links between research, policy, and politics. Here, the perspective on time differs from the historians' view. To start with, dominant development discourse—representing the view from the donor government or international agency, the ministry, consultancy firm, or even the giant corporations that in twenty-odd years have grown out of small non-gov-

---

41 Van Schendel 1998/2000.

ernmental organisations—is much more concerned with the future than with the past. The focus is on the present as a starting point for future improvement, as the hatchery of new processes rather than the outcome of old ones.[42] The central role of the concept of 'development' points to an evolutionary conception of world history in which humankind moves from backwardness to 'forwardness,' from 'non'-development to development, from tradition to modernity. Much in the tradition of evolutionary anthropologists of the nineteenth century, human societies are seen as progressing along a road (or possibly several parallel roads) towards development, some being more advanced than others. Of course, development scholars do not speak in terms of an evolution from barbarism to civilisation (as evolutionists did), and many have stopped using the markers of 'traditional' and 'modern,' but they do perceive the unrolling of time as 'progress' and a quest for human perfection, and in that sense, development is not just a destiny, but also a duty. In late twentieth-century development literature, and even more so in popular notions derived from it, Bangladeshi society shares with a handful of other societies (e.g. Ethiopia, Mozambique, Haiti), the distinction of bringing up the rear in the global march towards development.

In dominant development discourse in Bangladesh, time is linear.[43] It is anchored in the past, mediated by the present, and stretches outwards (or upwards) before us. But it does not necessarily 'flow' smoothly or evenly. The past is usually seen as 'underdevelopment'—understood as either the mere absence of the dynamism of progress or its active stunting by anti-development forces. In these studies, time begins to gather momentum during the most recent past, accelerating through the present and into a future heady with change. This has led to a remarkable feature: although large research funds have been lavished on development studies in Bangladesh over a long period, the history of development (let alone that of development policy) has been consistently neglected. Neither the emergence of a developmental state (which can be traced back more than a century), nor the developmental activities of non-governmental organisations in the past have drawn much attention.[44] Development studies in Bangladesh suffer from a distinct and, some say, intentional loss of memory. They are unwilling to confront the full effects of developmental activities in the past, especially over the longer run. It is here that the close links with policy and pol-

---

42 For a good example of this approach, see Sobhan 1982, Chapter 8.
43 For the sake of brevity, I have focused on the notions of time inherent in dominant varieties of development discourse in Bangladesh. This is not to say that there are not many different views within the 'development community,' often found in smaller non-governmental organisations; these views have had relatively little influence on concepts of time in Bangladesh.
44 For the emergence of a state development policy regarding the silk industry from the 1880s onwards, see Van Schendel 1995. Non-governmental organisations (NGOs), notably the Bengal Silk Committee and the Salvation Army, were precursors to state development activities.

itics emerge most clearly. In a sustained attack on the 'development discourse,' Escobar invites us to look at:

> ... development as a disaster of sorts which demands that the casualties be forgotten and dictates that the community that fails to develop is obsolescent. An entire structure of propaganda, erasure, and amnesia [is] orchestrated by science, government, and corporations which allow[s] the language of compensation as the only avenue of expression of outrage and injustice— and even compensation [is] precarious at best.[45]

Whether or not such orchestration can be shown to exist, the idea of obsolescence is useful when considering time in development discourse. The urgency which much of the development literature on Bangladesh radiates, and which is usually lacking in its historical counterpart, reflects a specific way of conceiving of time. The past is largely irrelevant, and it is no use crying over spilt milk. True, the 'allegedly beneficial and sanitised operations for the good of Mankind'[46] may have involved 'developmental' violence, and development interventions may have brought suffering for many, but these were just unfortunate and unintended side effects of a necessary cure intended to accelerate development.[47]

Acceleration can be combined with phases or periods projected into the future. 'Laggard' societies such as Bangladesh are depicted as being afflicted by social and cultural obstacles to development which developed societies have already overcome. Phases or stages of development (terms more commonly encountered than 'period' in this literature) were distinguished by the progressive removal of these obstacles. Explicit references to stages of development have, however, become less frequent with the weakening of modernisation theory; in the current development literature on Bangladesh, moreover, we encounter a diffuse sense of time rather than a clearly ordered one. The emphasis is on the present with selected references to a shallow past to indicate emerging development trends, a strong tendency to extrapolate into the future, and a seamless transition from conclusions to policy recommendations. The future is seen as providing great opportunities for further development; it is basically unstructured and malleable, open to strategic agendas, scenarios, and planning. Time becomes a *policy resource* which takes experience to appreciate; the seasoned development

---

45 Escobar 1995: 214.
46 Ibid.
47 The long silence about the effects of the Kaptai hydroelectric project, a national showpiece of development since the 1960s, is a case in point. Concern about the suffering it brought to the population of the Chittagong Hill Tracts, or the environmental degradation it caused, was not discussed in Bangladesh till the regional insurgency movement foregrounded it in recent years. Among a veritable avalanche of development studies, no balanced appraisal of the Kaptai project was ever commissioned. See e.g. Van Schendel, Mey and Dewan 2000.

expert may express 'impatience with those insufficiently patient; irritation with those who shoot too quickly from the hip.'[48] Unlike 'historians' time' (in which the past is a tool for charting the present, coping with shared experiences, and guiding and validating current behaviour) 'development time' uses the present as a tool for charting the future, creating shared experiences, and foreshadowing and validating urgent future action.

Embedded in the linear temporality of development studies in Bangladesh, however, are cyclical elements. First, Bangladesh is represented as strapped to the saddest seat in the global merry-go-round, forever chasing an elusive development by trying to catch up with societies which are always ahead. The literature never addresses the possibility of Bangladesh actually 'achieving' development, overtaking these societies, and becoming a model for them to emulate. Second, the development literature is closely tied to the short bureaucratic time horizons set by development projects, 'Five-Year Plans,' development decades, donor timetables, and ex-pat assignments. The 'project' is an important time unit in development studies; development time is, in a double sense, projected time, and the short life-cycles of development projects provide development time with a cyclical continuity.[49]

Modern Times in Bangladesh

Looking at the arenas of identity politics, historiography, and development discourse allows us to explore the complexity of the time regime in Bangladesh. The image of a faceless global time triumphantly invading a supine society is clearly not applicable. To start with, the social and economic pre-conditions for a 'metronomic society' never developed in Bangladesh because industrialisation remained very low.[50] Clock time, known in Bangladesh as long as in Europe or China, did not replace event time except in restricted areas of life and among special groups of Bangladeshis. Its influence has grown, but it is not easy to assess to what degree.

We have looked at two arenas in which global time was powerfully promoted. Academic historiography—with, in terms of Common Era periods, its construction of a state-centred national history—largely reflected the ambitions and interpretations of an emerging nationalist elite and was influential up to the

---

48 Wood 1994: 24, cf. 558.
49 Cf. Van Schendel 1995: 177-181.
50 Of all employed persons 10 years and above in 1990-91, 69 percent had as their main occupation 'agriculture, forestry and fishery' and only 14 percent 'production and transportation.' *Statistical Pocketbook* 1995: 120.

early 1970s. More recently, however, its influence has given way to that of development discourse, an equally ardent promoter of the global time regime but one which de-emphasises the past in favour of the present and the future. This shift was predicated upon the emergence of a new segment of the national elite, dependent on external development funds rather than state largesse.

It was in the third arena, identity politics, that the global time regime was actively opposed. Bengali Time and Islamic Time were imbued with new symbolic meaning and this ensured their widespread use. Rather than sliding into oblivion, they were instead politicised and presented as inalienable tradition. But Bengali Time, in particular, was continually adapted and manipulated. Although for many Bangladeshis, especially in the rural areas, the use of Bengali Time represents an unbroken practice, elite use of Bengali Time is better understood as an invented tradition.

In the second half of the twentieth century, the time regime in Bangladesh followed a course which differed from the predictions of time-convergence theory. Modern times in Bangladesh are multiple and dynamic. They show a dynamic which emanates as much from a steady expansion of a global time regime as from political struggles between secularists and Islamists in the country itself.

3. Plotting Time in Bali: Articulating Plurality

Henk Schulte Nordholt

A Complex of Calendars

People in Bali live in a variety of different time systems. This is illustrated by the calendar which hangs very visibly on the wall in most houses and is consulted regularly. Following the Western Gregorian time regime, this calendar is subdivided into twelve months, but it also contains a Hindu-Balinese, an Islamic, and a Chinese calendar, which seem to be of secondary importance. It would be wrong to conclude, however, that the Gregorian or national time regime dominates the other systems because there is yet another kind of time system prominently present on the calendar. This is the so-called Javanese-Balinese *wuku* calendar that regulates most of the ritual activities in Bali. Without this local calendar, many religious practices, which form a major part of Balinese identity, would be in total disarray.

In the following pages, I will discuss some aspects of the plurality of these time systems in Bali and the ways in which they are related to each other. Firstly, I will briefly introduce the *wuku* system and the various ways it can be contextualised within present-day Balinese society and the Indonesian nation-state. In the second place, I will argue that the Gregorian and *wuku* calendars in modern Bali do not simply exist side by side. Rather, they interact in a dynamic way, in the process of which political actors and economic factors play important roles as well.

The organization of time in terms of calendars enables people to think about time as duration, which makes thinking about the past possible. I will, in the third place, indicate how Balinese have conceptualised the relationship between past and present, and what kind of dilemmas they face as a result of the impact of modern historiography.

Finally, I intend to demonstrate that Balinese are not merely and passively subjected to the various time systems which are elaborated here. On the contrary, they have the capacity to interfere at, or in, certain moments in an attempt to regain control over their own destiny.

Quality and Duration

Whereas most calendar systems structure the duration of time in a quantitative way by adding a new year to the previous one, the Javanese-Balinese *wuku* system is primarily concerned with the specific quality of particular points in time.[1] It does not count, but rather, it qualifies.

Roughly summarised, the *wuku* system consists of three overlapping weeks of different lengths: one week of five, one week of six, and one week of seven days. This system results in a cycle of 210 days (5 x 6 x 7). The quality of a particular day is produced by the unique combination of the three overlapping week systems:

Day 1 to 210:	1	2	3	4	5	6	7	8	9	10	11	12	13	14	15	16 etc.
Five-day week:	1	2	3	4	5	1	2	3	4	5	1	2	3	4	5	6 etc.
Six-day week:	6	1	2	3	4	5	6	1	2	3	4	5	6	1	2	3 etc.
Seven-day week:	6	7	1	2	3	4	5	6	7	1	2	3	4	5	6	7 etc.

Each day is a combination of three different weekdays and has its own characteristics. Day one consists of the first day of the five-day week and the sixth day of both the six- and seven-day weeks. Day two has a completely different character because it consists of the second day of the five-day week, the first day of the six-day week, and the last day of the seven-day week.

This *wuku* system structures, by and large, the ritual calendar in Bali. It indicates at what days important temple festivals and other religious events should occur, and which days are or are not believed to be favourable for certain activities (Figure 3).

As such, every 210 days, the day of Galungan is celebrated. This is when, in most of Bali, deities and ancestors are invited by huge decorated bamboo poles to visit the adorned temple shrines which have been dedicated to them. Likewise, every 210 days, many temples celebrate their so-called *odalan*, or 'birthday.' Sometimes they do this in an elaborate manner, and at other times, more modestly.

1 Goris 1960; Kumar 1997: 139-158.

Figure 3
A Balinese calendar, 1998.

There are also special days for gods and 'things.' For instance, a specific day is dedicated to Saraswati, the goddess of writing and learning. Where I lived, lontar (dried palm-leaf books) were brought outside on Saraswati's day, and offerings were placed in front of them. Additionally, people were forbidden to read or write throughout the entire day. I was asked to bring my books as well, and to put the notebook I always had with me away for one day (which was indeed an unsettling experience for someone doing fieldwork). Another day, Tumpek Landep, is reserved for the maintenance of certain iron things, especially daggers (kris). Gradually, however, modern objects like motorcycles and cars have been included, and my Honda motor received a special cleansing as well.

The successful staging of important rituals, like weddings or cremations, depends very much on the ability to locate the right day. As such, it can happen—as frustrated tourists have noticed—that for weeks no cremation is held at all, whereas suddenly—just when their holiday on Bali is over—several extravagant cremations occur on one and the same day. Other activities, like planting rice plants or harvesting, are closely tied to favourable days as well.

In the old days, the *wuku* system was not the only time schedule in Bali. There was also, and still is, a Hindu-Balinese luni-solar system.[2] In contrast to the *wuku* calendar, this system is cumulative. At the end of every tenth month (*sasih kadasa*), the day of silence, or Nyepi, marks the beginning of a new Saka year.

The Saka year goes back to an ancient Indian calendar which runs 78 years behind the Gregorian system: In AD 2000 (or secularised CE), it is Saka 1922. In theory, thus, Balinese computers will encounter their millennium problem in AD 2078.

In ritual terms, the day of Nyepi is the opposite of Galungan day, when the gods and ancestors visit Bali. On the day of Nyepi, the island is haunted by demons, and people are obliged to stay inside in order to give the impression that Bali is empty. Unable to find any human beings, the demons will leave Bali.

Calendars in Context

In an influential article, Clifford Geertz concluded about the *wuku* calendar that '[Balinese] don't tell you what time it is; they tell you what kind of time it is.'[3] While elegantly phrased and very quotable—something most of his followers as well as his outspoken and longstanding critics have yet to achieve—it is only to some extent true.

By focussing almost exclusively on the *wuku* system as the only dominant time system in Bali, and imprisoned by his own analytical concept that a cultural system governs people's actions to a large extent, Geertz was led to the conclusion that Balinese social life 'lacks climax because it takes place in a motionless present, a vectorless now, or, equally true, Balinese time lacks motion because Balinese social life lacks climax.'[4]

In a response to Geertz, both Maurice Bloch[5] and Leo Howe[6] have argued from different perspectives that 'the' Balinese have, in addition to 'cyclical'

---

2 According to this calendar, a year consists of 12 months of 29 or 30 days, or in total, 354 to 356 days. After 30 months, an extra month of 30 days is added in order to keep the system in line with the luni-solar cycles. Most temples in the south Balinese rice plain follow the ritual wuku calendar of 210 days, which goes back to the eleventh and twelfth centuries when Bali was influenced by East-Javanese court culture. However, a number of Balinese mountain temples tend to follow the luni-solar calendar, which dates from a much earlier period during which Indian influences reached Bali. A major reason why these older influences have survived is the fact that the mountain temples still play an important role in the maintainance of the agricultural cycle. These temples celebrate their major rituals in September-October, when the harvest would be over. It would not have made any sense if these temples had followed a non-agricultural cycle of 210 days.
3 Geertz 1966: 47.
4 Geertz 1966: 61.
5 Bloch 1977.
6 Howe 1981.

notions of time, an awareness of linear time as well, either because this is a universal phenomenon (Bloch), or because it is part of Balinese culture (Howe).

Recently, Mark Hobart has reopened the debate on Balinese notions of time by accusing anthropologists of denying the Balinese a sense of history. In his lengthy and, in my opinion, rather disappointing article,[7] Hobart concludes that Balinese culture is not a bounded whole, and that we should concentrate on Balinese practices which are diverse and vary situationally. I cannot say that I am really astonished by these findings. What is worse, after 50 pages of academic oratory, what Hobart calls the 'missing subject'—that is, Balinese representations of time and history—is still missing.

Apart from the fact that Balinese culture cannot be seen as a bounded whole and that concepts should be seen in contextualised practices, Bali itself is not an isolated place. Like some of his colleagues, Hobart seems to underestimate the extent to which Bali has become part of larger economic and political worlds, in which Balinese live and act as well. This can best be illustrated by pointing to a tragic coincidence: When Geertz wrote in 1966 that the timeless Bali lacked climax, about 80,000 Balinese had just been killed in the wake of the abortive coup in Jakarta.

Since the beginning of the twentieth century, Bali has become part of the larger administrative structures of the colonial state, as well as of the capitalist world market. A consequence of these processes has been the encounter with a colonial time regime. Like elsewhere in the world, the spread of railway networks stimulated the establishment of a uniform time system.[8] From the nineteenth century onwards, the Dutch had introduced a very localised time system in the Indies. Every town in Java had its own time, which was separately related to Greenwich Mean Time (GMT). The purpose of the new time system with regard to a colonial labour regime, how it was communicated as well as the confusion it caused, is nicely illustrated in P. Pieters' language course for the colonial freshman:

- Do you know what time it is?
- ...
- Look at your watch.
- It does not run parallel with the city clock.
- I haven't synchronised it.
- ...
- Do the villagers have a clock?

---

7 Hobart 1997. The article is a mixture of his ongoing and eventually rather predictable crusade against almost everything Clifford Geertz ever wrote, and his habit to demonstrate how witty and philosophically well-grounded his opinions actually are.
8 See Cribb 1998 and also Goudsblom, in this volume.

- Some wealthy villagers have a watch.
- But a commoner does not posses a clock.
- How do they know what time it is?
- ...
- Every hour a wooden alarm clock (*kentongan*) is sounded.
- By whom?
- The clock is sounded by the guard of the district head.
- When do working hours start?
- Labourers work from six o'clock in the morning till six o'clock in the evening.[9]

It was only after the introduction of six time zones in the Netherlands Indies in 1935 that efficient time schedules for the departures and arrivals of trains could be made. Bali belonged, together with Java, to the third zone (GMT + 7'30").

Soon after the military expeditions of 1906-08, south Bali was incorporated into the colonial state. The Dutch administrators had transformed one of the south Balinese royal residences, Denpasar, into a new administrative centre. The old royal palace and adjacent noble dwellings had been demolished, and new European offices and houses were built at the same spot in order to make the new colonial rule manifest. The European houses became a desirable model for Balinese who aspired to a modern lifestyle, and they named their new houses after the administrative buildings of the new rulers, *rumah kantor* (office house).

Significantly, the Dutch located a huge clock at the central crossroad of Denpasar. In the old days, this spot had been one of the centres of the realm where important cleansing rituals were held, but now the clock communicated that times had changed. Clock-time discipline was gradually imposed on a growing number of Balinese subjects.

Following central clock-time, alarm clocks and bells reminded children all over Bali of the obligatory school hours, while a small number of Balinese clerks were obliged to be present at their office at seven o'clock sharp. The majority of the Balinese was still not (yet) captured by the new time regime, but they had at least heard, and wondered, about it. Little by little, the new state time gained ground in Balinese society.

It would be a mistake to assume that the local Balinese time systems remained untouched under the new regime. Since the colonial administration was increasingly involved in the codification of local customary rules (*adat*) as part of an effort to freeze Balinese society in a rigid set of traditions, the ritual calendar was standardised as well.[10]

---

9 Pieters n.d.: 240-244.
10 Personal communication of David Stuart-Fox. The article by Goris (1960) was probably part of the colonial effort to 'restore' the ritual calendar. On the invention of tradition in colonial Bali, see Schulte Nordholt 1994.

In former days, differences between various Balinese kingdoms had occurred in the timing of rituals. In 1828, for instance, the king of Badung had suspended the ritual calendar because of a devastating smallpox epidemic. People were ordered to stay inside, and by eliminating the progress of time, death could be averted.[11] Likewise, Goris noted that in the former kingdom of Bangli, the day of Nyepi occurred on a different day than in the neighbouring areas.[12] By identifying these differences as anomalies, the Dutch, and later, the Indonesian administration, synchronised the ritual calendar.

The colonial state did not only 'restore' the local ritual calendar to its 'proper order,' it also introduced its own 'holy days,' like the Queen's birthday, which was celebrated on August 31. Despite the Dutch efforts to emphasise the traditional nature of Balinese culture, Balinese themselves had a keen sense of modern times. An intriguing new word was seen on recently built or restored temples, *kelar* (or 'klaar' in Dutch). This was followed by a date indicating when the building was finished, such as '12-4-1928.'

After Indonesia gained independence in 1945, Bali was not only relocated within a set of three reorganised time zones (in 1950, in the central zone [GMT + 8']; in 1983, in the western zone [GMT +7']; and in 1993, again but without Java, in the central zone [GMT + 8'] (Cribb 1998), it also experienced a series of national commemorative days, of which Independence Day on August 17 was the most important. It was, however, not just one-way traffic of national celebrations which were imposed by the centre upon the regions. As a result of the New Order policy to respect the various religions in Indonesia—and, at the same time, to downplay the fact that Muslims form an overwhelming majority—the day of Nyepi, marking the Balinese new year, became a national holiday in the early 1980s. Many Balinese were of the opinion that the day of Galungan, when the gods and ancestors visit the island, was a better occasion to put Bali on the national calendar, but since that day is part of the *wuku* calendar and occurs every 210 days, it does not fit into the Gregorian timetable of the state. Therefore the whole nation celebrates every year the fact that Bali is haunted by demons and looks like a deserted place.[13]

The huge clock on the central crossroad of Denpasar that marked colonial discipline, has been replaced by a statue of a four-faced god, but that did not mean that state agencies had mitigated their efforts to discipline Balinese society. Although under the New Order regime popular participation in Independence Day activities had gradually decreased, the state encouraged the elaborate staging of religious events, like the enormous island-wide cleansing ritual, Ekadasarudra, in 1979, and the somewhat smaller Panca Wali Krama, which is held every

---

11 Lovric 1987: 130-131.
12 Goris 1960.
13 In 1998, Nyepi coincided with March 29, in 1999 with March 18, in 2000 with April 5, and in 2001 with March 25. See http://www.egroups.com/list/bacn.

ten years. Because President Suharto and other state authorities witnessed the finishing of the Ekadasarudra in 1979, the ritual also symbolised the harmonious relationship between loyal subjects and their benevolent protective patrons, and the subordination of Bali to the New Order.[14]

In modern Bali, people live in a plurality of time systems, which imply various ways of remembering. While a schoolchild memorises that the National Awakening of Indonesia occurred on May 20, 1908 (when the organisation Budi Utomo was established on Java), the name of her grandfather, I Wayan Gejer (or 'trembling') is a reference to the time at which he was born, during a devastating earthquake in 1917.

While Putu Oka reads about the economic crisis in Indonesia in the December 17, 1997, issue of the local newspaper, *Bali Post*, he will tell me that something in the village happened five days after the *odalan* of temple X. And his wife will inform me that she is making a small offering because it is now day so-and-so on the *wuku* calendar, which is her daughter's *oton*, or 'birthday,' and that her daughter wants to celebrate her other birthday, according to which she becomes 16 years old, in a modern way. I am asked to make pictures of her party where there is a big cake with 16 candles and modern-dressed guests who sing that 'she is a jolly good fellow.' Both the *oton* and the modern birthday are part of her life. Indeed, depending on the context, Balinese will give various answers when you ask them what (kind of) time it is.

Capitalist Time – Ritual Time: The Yellow Mercedes

The incorporation of Bali into a wider state system also connects the island with a capitalist economy. One of the most valuable commodities of Bali is its tourist culture, which is consumed by thousands of visitors every year.[15] The fact that their lively Hindu culture is the core of a profitable tourist business increased the self-awareness of many Balinese. It made them realise that their small island is a unique and, in terms of tourism, crucial place in the Indonesian archipelago. It also made an accommodation between capitalist time and ritual time imperative.

The Brahman family of Y consists of about 250 persons, subdivided into 70 households. In the old days, the core house of the family was the seat of royal priests, but after independence, things changed. Political parties entered the regional arena, and part of the Brahman family became active members of the communist party, PKI. In 1965-1966, after the abortive coup in Jakarta, the PKI was annihilated, and about 80,000 people were killed in Bali. The Brahman fam-

---

14 Schulte Nordholt 1991a.
15 Picard 1996.

ily was also haunted. Several of its members disappeared, while others were socially ostracised and excluded from government jobs for decades following.

When, during the late 1970s, tourism started to boom in Bali, some of them tried their luck in the tourist business and succeeded. They opened a hotel, a restaurant, another hotel, and they became involved in a travel agency, money changing, and accountancy. Gradually, they improved their relationships with the regional government—which was now led by a Brahman to whom they were vaguely related—and this gave them better access to official permits and the like. Ideally, they could have established an integrated tourist enterprise, but they did not. Instead of running a family company, the brothers and cousins each own and mind their own respective businesses. As one of the family members said:

> It is better if we do it this way; it is more professional. It would be a mess if we would have shared the responsibility, because in the end nobody would be responsible. It would be like a government cooperative, and the family would split because of all the complicated quarrels we get.

The running of their companies is an individual responsibility and requires a management of capitalist time, which is oriented toward the future in terms of investments, interest rates, and MBA studies in the USA by the junior members of the family. At the same time, however, the family invests a lot of time and money in ritual activities in order to keep the 70 households together, and many of these rituals are organised around ancestral shrines. In order to face the future, they have to revive the past. The family members see each other on a very regular basis: during the *odalan* of various family temples, at cremations of senior relatives, and at the elaborated celebration of marriages. There is a strong moral obligation to attend these time-consuming festivities:

> If you miss once in a while a ritual, nobody bothers, but if you are twice absent, it is noticed. And if you do not show up for a third time, a senior family member will come and visit you and tell you that such an improper behaviour is not at all appreciated and cannot be tolerated.

Rituals tend to become more elaborate as the wealth of the family increases. But since both male and female members of the family are involved in their business activities, they lack the time for the lengthy preparations of various rituals. Therefore, they order complete sets of offerings and decorations from commercial Brahman companies specialised in the catering of rituals. An extra asset at these occasions is an original ritual instrument from India, while the co-ordination of a long cremation procession is done through mobile phones. Another effort to organise rituals more efficiently is trying to plan the proper (*dewasa*) days for a ritual as much as possible on Sundays. Ritual time negotiated by capitalist interests.

We do try to make these rituals more efficient, but if it is necessary, I will attend a ritual for a couple of days. During these rituals we see each other. And of course we talk, we inform each other about business opportunities, so the family is a good network as well. But what is more important, these rituals help us to remember who we are.

Time is indeed money, but it is only through ritual that money begets meaning. As a colleague of mine in Bali said: 'Rich Balinese dream of a yellow Mercedes.' It is the vehicle of capitalism and the colour of both Balinese royalty as well as Golkar, the ruling party under the New Order. But since the fall of Suharto in May 1998, Golkar is no longer a safe investment in a secure future.

Capitalist Time – Ritual Time: The Governor's Defeat

One of the things that worries regional government authorities in Bali is the resilience of local customary practices that can be used by the population as instruments against the penetration of unwanted state interference.[16]

Since the mid-1980s, large-scale investments in the Balinese tourist sector have increased, and especially in recent years, strong Jakarta-based investors have literally gained ground in Bali. The governor of Bali, Ida Bagus Oka (1988-1998), has facilitated this process. Instead of protecting the interests of the island and its cultural and religious integrity, he was seen as a corrupt agent of external forces selling Bali to foreign capital. For this he was heavily criticised in the Bali Post and ridiculed in cartoons as Ida Bagus 'Okay.'

In this respect, a hotly debated case regarded the establishment of the exclusive Bali Nirwana Resort at the south-west coast of Bali in 1994. Besides the fact that 212 hectares of fertile rice lands were converted into a golf green, the resort was, according to many Balinese, located too close to the sacred sea temple of Tanah Lot. Such a brutal penetration of foreign capital, insulting a proper ritual process, caused fierce protests from all over Bali. Eventually, a sort of compromise was reached: a neutral corridor of two kilometres would keep the resort at a decent distance from the temple. By and large, the protesters saw the compromise as a defeat and pointed to the bad role played by the governor.

In 1997, a similar case occurred, this time with a damaging result for the same governor. In October, the Bali Post reported that at Padanggalak, the coastline of the village of Kesiman east of Denpasar, a large tourist resort was being planned and that building activities had already begun. The owners of the project were based in Jakarta, but the actual building was done by a local firm. The newspaper also revealed the involvement of the governor. Ida Bagus Oka had

---

16 See Schulte Nordholt 1991a; see also Warren 1993.

given his consent to the plan, and it turned out that his own office was involved in the tender.

The beach where the new resort was planned was an important ritual site for both the village of Kesiman and thousands of inhabitants of east Denpasar. It is the place where regular cleansing rituals (*malasti*) were held and where the final post-cremation purifications took place. Religious events scheduled by the ritual calendar were now threatened by the merciless rhythm of time-is-money capitalism.

Although the semi-government agency for Hindu affairs, the Parisada Hindu Dharma, had also given its consent to the plan, a broad coalition of students, intellectuals, and local people from Kesiman protested against the project. The protest was given voice by the *Bali Post*, the still independent newspaper which carefully maintains 'the spirit of 1945,' i.e. the values of the national revolution.[17]

Within the village of Kesiman, a powerful alliance was formed between villagers, customary (*adat*) leaders, and the local nobility. They declared that the voice of the village *kulkul*, the wooden alarm clock, would eventually prove to be stronger than a government letter.

Support also came from the well-known regional historian Ketut Subandi (see below), who argued that the beach was a sacred site because, in ancient times, an ancestor of Ida Bagus Oka had rested there during one of his journeys. Additional support came from Megawati Sukarnoputri, the daughter of former President Sukarno and leader of the nationalist opposition party PDI-P, who is extremely popular in Bali.

Resistance in the village was co-ordinated by Anak Agung Kusuma Wardana. He presented himself both as a villager who obeyed the local *adat* and as a representative of the local nobility who was willing to defend his village against outside attacks. Moreover, as a member of the ruling Golkar party, he had a seat in the regional parliament where he put the Padanggalak case on the agenda. It was, however, not in parliament, where the issue was decided, but in the village of Kesiman.

Because the governor originated from the village of Kesiman, the opposition decided to hit him locally. After a meeting of village notables in the residence of Anak Agung Kusuma Wardana, the village council threatened to ostracise Ida Bagus Oka from the village. In terms of everyday normal life, this had perhaps not many consequences; in ritual terms, however, it would be a disaster for the governor. It meant that the village would not allow him to participate in village rituals and, what is worse, would eventually boycott his cremation, which would prevent the proper continuation of the journey of his soul.

When he heard about the village verdict, the governor immediately declared that nothing had actually yet been decided, and that the final decision about the

---

17 Warren 1994.

project was, of course, in the hands of the village council. Apparently, Ida Bagus Oka did not want a clash at the end of his last term as governor which might have jeopardised both the future of his soul as well as an appointment in Jakarta—which eventually materialised in May 1998, when he became a junior minister *cum* chairman of the National Family Planning Board.

With remarkable speed, all the building activities were cancelled, and by the end of November the beach of Padanggalak had been restored to its former shape. In the aftermath, the Golkar party tried to discipline its unruly member of parliament, Anak Agung Kusuma Wardana, but failed. The effort itself caused new protests which further undermined the legitimacy of the party, already in decline as a result of the nation-wide protests against the Suharto regime. Finally, Anak Agung left the party in July 1998 because he felt that as a member of Golkar, he could no longer serve the aspirations of the people.

The resistance in the Padanggalak case concerned the continuation of local ritual practices which were organised by the ritual calendar. Firmly based on a ritual issue, a broad and successful coalition was mobilised against a capitalist intrusion which was supported by Jakarta-based institutions and a corrupt governor. This case serves to illustrate that by ostracising the governor in a ritual sense, the opponents of the governor were able to manipulate (or qualify) his future fate. Defending the ritual integrity of the island, the protest was also a manifestation of the concern with the future of Bali.

Coinciding the Present and the Past

In the past Balinese had systems at their disposal to calculate both the quality of particular moments in time, as well as to measure the agricultural cycle. Moreover, by using both the *wuku* system and, more particularly, the Saka calendar, they were able to date certain moments in a chronological order. This is illustrated by the existence of long lists of so-called chronograms, or *Candrasengkala*, commemorating eventful, and often disastrous, moments from the past.

These chronograms often consist of codes in which, in reversed order, certain words refer to a particular year. For example the code indicating the defeat of Blambangan in East Java is '*nora tinghal bhuta tunggal*,' which stands for '0 2 5 1.' In reversed order, we see the Saka year 1520, which is AD 1598 (i.e. 1520 plus 78 years). Seen from a conventional western historiographic perspective, this date is correct, but many others are rather unreliable.[18]

---

18 I would like to thank Hans Hägerdal for generously providing me with his list of chronograms. For a discussion, without firm conclusions, about the chronology of Balinese history in the seventeenth century, see the exchange between Creese (1991, 1995) and Hägerdal (1995a, 1995b).

What really concerns the majority of the Balinese, however, are not series of precise dates, but genealogical narratives, or *babad*, in which the ancestral origin of their family and the relationship between ancestors and descendants is revealed. Elsewhere, I have analysed such a genealogical narrative of a south Balinese dynasty within its political context.[19] The *babad* is not an exercise in recording past events in a factual and chronological order. Instead, it organises and explains, in terms of time and place, the hierarchical order which emanated from the dynastic centre towards the periphery of the realm. It was, in the very first place, a state-building text. The narrative emphasises continuity and elaborates moments of crisis when the continuity of the dynasty was at stake but dynastic leaders managed to maintain their precedence.

Despite the availability of series of chronograms and the genealogical continuity which is suggested by many *babad*, Balinese representations of the past are very episodic. For example, certain episodes in the past are very densely described, while other periods are almost empty. Most of the eighteenth century, the first half of the nineteenth century, and a good deal of the colonial period are 'missing' in Balinese historiography because nothing much happened—that is, no dramatic nor memorable turning points occurred. Consequently, if one talks with older Balinese members of the nobility about their past, they will generally relate the following (undated) episodes, which highlight turning points in their ancestral history and inform them about their own identity: 1) when their ancestors came from Java to Bali [referring to events between AD 1300 and 1600, but many express this by saying 'when I left Java,' and 'when I arrived in Bali'], and when the principles of the hierarchical order were established; 2) the disintegration of the Balinese realm of Gelgel, and the establishment of noble centres in other parts of the island, i.e. the reshuffling of the hierarchical order [in the second part of the seventeenth century]; 3) the involvement of many families in the big warfare in connection with increasing colonial expansion [at the end of the nineteenth century], resulting in defeat or accommodation at the time of the Dutch military expeditions [between 1894 and 1908]; 4) the national revolution [1945-1950], when the position of many noble families who had served in the colonial administration was challenged by new revolutionary forces.[20]

In order to describe, and by that to contemplate, the causes and forces that led to the destruction of fragile hierarchies, Balinese used another genre of texts, the so-called *uwug*. In contrast to the prose texts of the *babad*, which were simply read aloud, *uwug* were poems that had to be sung. And the perfection of its composition reflected the truth of the text. It could be argued that Balinese have, or

---

19 Schulte Nordholt 1992.
20 The mass killings of 1965-1966 form yet another dramatic episode in Balinese history. For 30 years, it was a political taboo to talk about these events, but perhaps the new situation in Indonesia will provide openings for an historical confrontation with that tragedy. Cf. Robinson 1995.

had, a cyclical notion of time because, in many instances, reference is made to the arrival of the *kaliyuga*, the time of confusion, which is followed by the *pralaya*, or total destruction. This may be the case, but the same texts hardly ever refer explicitly to the other side of the wheel—the Golden Age, or *krtayuga*. The wheel of time is, in other words, not turning because Balinese histories seem to have more affinity with disaster.

The episodic way of telling about the past is not a unique feature of Bali, for it reflects a very common way of representing history. One may, in this respect, wonder how exceptional modern Western professional historians really are, with their peculiar interest in strange subjects like everyday life and daily routines of a distant past. Like a lot of people in the Western world, many Balinese do not see why it would be important to record non-eventful histories which have, moreover, no immediate relevance for their own lives either. If the study of history, in terms of a series of chronological events, did not seem to have much relevance for Balinese, how should we understand the role of the coded dates mentioned above?

Gregory Bateson has argued that in Bali, 'the past provides not the cause of the present but the pattern on which the present should be modelled.'[21] Consequently, the past should not be seen as a distant world which differs fundamentally from the present, but as near and familiar. It is, in other words, not the quantity, but the quality of time that matters.

In an important article, on which I will lean rather heavily here, Adrian Vickers demonstrates convincingly that Balinese notions of time and fate have produced a logic that has patterned their representation of historical events.[22] In particular, he looked at so-called *pangéling-éling*, commemorative notes which were attached as a colophon to a variety of Balinese texts about romance, warfare, and court culture of the past. The attachments indicate when the writing of the text was finished and refer often to a dramatic event, like a war or natural disaster, which had occurred in Bali at the time of writing. As such, the *pangéling-éling* in the colophon made the content of the text coincide with an event in the outside world. By making this connection, authors contextualised the texts in the present and located dramatic events within a larger pattern of texts.[23]

In Balinese discourse, thus, history is not gradually fading away into a distant and unfamiliar past, because the past must be kept in the present, and the present has to be in line with the past.

---

21 Bateson 1970: 135.
22 Vickers 1990.
23 Vickers 1990: 176-177.

Making Memory

Just like the African Azande,[24] many Balinese are convinced that things do not happen by sheer accident. In their eyes, such a non-explanation which defines the inexplicable as accidental reflects a poverty in Western culture. Instead, Balinese are convinced that many events are a fated combination of divine and demonic forces:

> Balinese describe their lives as a set of overdetermined moments. On a personal level such moments are birth, illness, marriages, and death. In village and state terms these are or were epidemics, wars, state rituals, and other similar critical happenings. The significance of these events can be gauged from their correspondence to textual events. [...] 'Events' or the 'nodes' of time are points at which elements of 'fate' come together. Texts are meant to give a sense of the arrangements or order of 'fate' and of how much an individual may act.[25]

Balinese believe that the past causes effects in the present because the distance between the two can be minimalised by making things to coincide. However, the coincidence created by linking the past through a text to an event in the present does not simply explain the event and reveal its 'fate' in a direct way. Instead, it opens the possibility of endless speculative interpretations.

The making of a coincidence is not merely intended to grasp the meaning of a 'fateful event,' it can also be an effort to influence the quality of the course of time:

> Time is here also a destructive deity, Kala, to be held at bay by propitiatory rituals or powerful acts, and thus turned into one of its less malevolent forms, the god Siwa. The quality of divine manifestation is a quality of time.[26]

It is not only through texts and rituals that Balinese have tried to influence the quality of time. In the following example, architecture and the organisation of space were also used to keep the past and the present in line with each other in order to effectuate historical change.

During my fieldwork, I lived in the south Balinese village Blahkiuh, and only gradually did I start to understand the whereabouts of its former political centre.[27] Under colonial rule, the village had been the residence of Gusti Mayun, a

---

24 Evans-Pritchard 1937.
25 Vickers 1990: 177.
26 Vickers 1990: 170.
27 For a more detailed account, see Schulte Nordholt 1991b.

member of the former regional dynasty. He had been appointed by the Dutch as a district administrator and had wanted to restore the authority of the former dynasty.

The residence of Gusti Mayun was a Western villa (*kantor*) which materialised the message that the administrator was a modern man. After some time, I learned that the house was part of a much larger project designed by Gusti Mayun. He had also created a new crossroad which redirected the centre of the village away from the old village temple and toward his new villa located at the north-east side of the new crossroad. The crossroad—where at noon the dangerous god Kala dwelt, and where, following the *wuku* calendar, important cleansing rituals were held—was soon overshadowed by a fast-growing banyan tree, which dominated the centre and could be seen from miles away. Southeast of the crossroad and in the shade of this holy tree, a lively market was held, which attracted many people to the centre.

At the south-west side of the crossroad, a respected Brahman family was housed. This was not the residence of a Brahman priest, like it used to be in former days when king and priest were never far away from each other. Instead, several Brahmans served as officials in the local administration.

The most important building was a temple located at the north-west side of the crossroad, facing the holy mountains. This temple concentrated various important functions. It had a large wooden alarm clock (*kulkul*), which became the 'voice' of the regional administrator, while a special shrine was dedicated for facilitating irrigation and fostering fertility. As a result, the authority of Gusti Mayun was ritually enhanced and associated with the well-being of the population.

There was, at the same time, a dimension which puzzled me for a long time. A man who 'possessed' the story of the temple shared it with me:

> A long time ago, a man from Blahkiuh made a journey to the capital of the kingdom of Mengwi during the reign of the first great king of Mengwi. Halfway, he stayed overnight in a village hall, and during his sleep he received a divine message to look in the ceiling of the hall for a letter which he had to deliver to the king. When he woke up, he saw the letter and brought it to the king. The letter was a divine order to build a temple in front of the market. However, in Mengwi there was already such a temple, and therefore, he decided that the temple had to be built in Blahkiuh. And that is why, to the present day, there is a temple on this spot in Blahkiuh.

According to this story, I calculated that the temple must have been built in the first half of the eighteenth century. Several months later, I talked with the same man who had 'given' me the story, who then casually said 'Oh ya, that happened in 1928, when my father was supervising the building of the temple.' Indeed, an inscription mentions 1928 as the date when the building was finished.

How do we reconcile these two pieces of contradicting information from one and the same informant? It took me a long time to understand that we there are here two different concepts involved. There is the fact that the temple was built in 1928, a fact that people happen to know (*uning*), but that has in itself no particular significance. What is far more important is the fact that the temple had to commemorate (*éling*) the linkage between the authority of the old dynasty and the royal ambitions of Gusti Mayun, the modern administrator.

In connection with the temple, he also (re-)composed the genealogical narrative of the dynasty. In a way, the new temple with its old story was an architectural *pangéling-éling*, diminishing the distance between the past and the present by connecting the days of the dynasty in a meaningful way within a new colonial context.

Encounters with Modern Historiography

Since Bali has become part of the Indonesian nation-state, a national identity has been communicated through mass media, schools, and historiography. Under the New Order regime of President Suharto (1967-1998), a developmental authoritarianism was established in order to achieve rapid economic growth in combination with political stability. The centre of the state was seen as the only dynamic operator of a controlled process which would lead Indonesia to a new era of progress and prosperity. This centralist approach resulted in a very centralist and eschatological historiography: national leaders dominated the making of modern Indonesia, which unfolded itself in a linear process. In the official Indonesian historiography, regional histories were marginalised, and if mention was made of local histories, they had to fit into the larger pattern of the nation's biography.[28] The marginal position of Balinese history is, for instance, reflected by the fact that only recently two Balinese have been added to the pantheon of more than a hundred national heroes: in 1975, the revolutionary leader Gusti Ngurah Rai (d.1946), and in 1993, the north Balinese leader Gusti Ketut Jelantik (d.1849), both of whom lost their lives while fighting the Dutch.[29]

A similar attitude existed under the New Order with regard to regional cultures, which were seen as static entities that needed to be protected and improved by agents from the centre. Such a perspective reinforced the idea that regional histories had no dynamics of their own. Local practices were subordinated to interests of the centre, which wanted to make its national developmental polity a success ('*sukseskan pembangunan national*').

---

28 Creese 1997.
29 Schreiner 1997.

The new national historiography emphasised the authority of (colonial) state archives as repositories of true facts expressed in concrete dates. As such, national historiography added another genre to Balinese representations of the past, and this received the Indonesian word for history, *sejarah*. Ironically, historians from the local university who conducted government-sponsored research projects to record the regional history relied primarily on Dutch archival materials (which included a very colonial perspective) because they considered their own Balinese sources to be too unreliable.

An effect of the national historiography was that more and more dates tended to be incorporated into new versions of Balinese representations of the past. Gradually, the idea gained ground that texts without dates are less reliable. Soon, however, Balinese historians faced an unforeseen problem, as expressed in the following anecdote:

> In 1989, I finally succeeded in getting access to the house of Anak Agung X. He belongs to a prominent Balinese dynasty and made an impressive career in Indonesian politics and diplomacy. He is also a Western-trained historian who obtained his PhD thesis abroad. When he retired, he came back to Bali and devoted his time to the ritual obligations of his family towards 'his people' and the writing of his family history cum autobiography.
>
> Part of his mission is the revival of the old court culture (which would be why I found myself surrounded by submissive servants when I met him).
>
> Yes, he was interested in seeing me, a colleague historian, and after the polite opening of the conversation in which he asked how Queen Beatrix, her husband, and his other colleagues were doing, he soon came to the point. Offering me a whisky, he told me that he was investigating his family history. He had consulted his genealogical narrative (*babad*) and other sources. The problem was that according to the *babad* and to calculations of the number of generations, his ancestors had arrived in Bali about AD 1500. However, this should have been AD 1343. That was the date given by the most prestigious Balinese *babad*, the *Babad Dalem*, which indicated the arrival of the Javanese nobility in Bali and the establishment of a new hierarchical order. He faced a gap of more than 150 years, which he had to fill with a decent ancestry, otherwise he was no longer connected with the legitimazing source of his elevated status. Such was the dilemma of a Western-trained historian who was, at the same time, a loyal descendant of an old royal family.

Helen Creese has rightly observed that in Bali 'totalising concepts such as national culture and history [...] inevitably give rise to resistance and counter-hegemonic practices.'[30] Besides efforts to impose a national historiography, or

---

30 Creese 1997: 24.

to reconcile Western historiography and local practices in order to anchor one's identity in a familiar past, there also emerged a popular historiography in Bali. This genre was, in the first place, embodied by a former head of police and son of a temple priest, I Ketut Subandi. Since the late 1970s, he had been the spokesman of commoner groups who were also in search of their ancestry.

In contrast to the nobility, commoners were not used to having elaborate genealogical narratives. In most cases, people had only a vague notion about their ancestry. This changed considerably from the late colonial period onwards, when diffuse descent groups were gradually moulded into larger temple networks and formal organisations, and these started to question the supreme status of the high-caste Brahmans. Under the leadership of Ketut Subandi, various Pasek commoner groups—once an honorific title for village notables—were transformed into an island-wide kinship organisation. A streamlined genealogy connected various groups with each other and with an ancestry that had arrived in Bali much earlier than the ancestors of the nobility. Hence, Pasek actually had a higher status than Brahmans.

In order to demonstrate these connections, Subandi published a book in which he also gave a series of exact dates—starting with Saka 11 (AD 89), when snake-gods gave Bali its holy volcano—which were derived from Balinese chronograms.[31] Since he has become a temple priest himself and gained prestige as a popular historian, he has changed his name to Jero Mangku Ketut Gede Subandi. He also started to write a column in the daily newspaper, Bali Post, in which he provides people who are searching for their ancestors with ready-made genealogies upon which they can base their identity. As such, he succeeded in writing an alternative popular history of Bali which offers everybody his much-needed ancestor.

Articulating Times

In his inspiring essay in this volume, Goudsblom seems to suggest that the modern capitalist time regime eventually dominates other, previous time systems. What I have tried to show in this chapter is that such a subordination of local systems by a new outside regime did not occur in Bali. At the same time, we should not look at Balinese calendars in isolation from other time systems. Instead, we have seen a complex articulation between a state cum capitalist time regime and a local ritual and calendar system, through which strategically important counter-identities can successfully be communicated and realised. In this respect, it is important to note that the local ritual calendar is not static, but has been, and is still being, modernised. The reproduction of the ritual time system is the result

---

31 Subandi 1982.

of a complex interaction between institutions at the state level, and regional agents who define the Hindu identity of the Balinese, as well as its profitability, in terms of tourism.

What the outcome of the changes in the post-Suharto era will be is not yet clear. Like elsewhere in Indonesia, there is a strong revival of regionalism, which will, without a doubt, affect both Balinese historiography and ritual practices. Bali is a relatively rich area, and many Balinese want more autonomy vis-á-vis the weakened centre of the state in order to keep profits for themselves. There is also an increased concern for what is seen as the rising tide of Islam in Indonesia. In this respect, the mass support for Megawati during the PDI-P congress in October 1998 in Bali was also a mass demonstration against Islam. When, shortly after the PDI-P congress, a Muslim cabinet minister, Sjeafuddin, told the press that it was inconceivable that Megawati could ever become president of Indonesia since she had prayed in a Hindu temple where she had worshipped idols, no less than 50,000 Balinese gathered in Denpasar to demand his resignation.[32]

Given the present context of the rise of both Islam and regionalism, Balinese will use all the means they have at their disposal to ground their identity in their own ritual calendar and history in order to 'sukseskan' their own future.

---

32 It is not the first time that Balinese are faced with the rising tide of Islam. In 1635, the king of Bali refused to accept a request from the ruler of Makassar to convert to Islam, and answered the envoys from Makassar that if their ruler had plans to conquer Bali, an army of 70,000 men armed with lances greased with pork fat would be ready to fight him. See Wessels 1923.

4

The Quest for Universal Time: Periodising the Past in Japanese (and German) Historiography

Sebastian Conrad

The temporal order of past events is one of the central aspects of the construction of historical meaning. In their accounts of the past, historians go beyond the uniformity of calendar time to structure their historical narratives. They rely on a concept of time that includes discontinuities and turning points, revolutions, phases of acceleration or stagnation, and 'long' and 'short' centuries. These forms of periodisation are subject to change and reveal not only the guiding principles and interests, but also the unspoken assumptions on which all interpretation is based. The narrative structuring of time permits the transcendence of a simple chronological order and bestows significance and meaning upon the mass of inchoate facts. All narratives that turn disconnected events of the past into a coherent historical plot—the history of the nation, for example—rely on an order of time that serves as a guarantor of continuity and synthesis.

In what follows, therefore, I will not be so much concerned with a specific cultural concept of time; this is not a contribution to an historical anthropology of an indigenous Japanese or German approach towards temporality. Rather than speculate on the impact of Confucian, or possibly 'native', thought on a Japanese understanding of time, I will focus on the function of 'time' in historical discourse. Thus, time is not taken here as an *a priori* category of human thought determining the relationship between the historian and his object of study in culturally specific ways. I am not, in other words, dealing with questions of epistemology. Instead, I am concerned with aspects of historical representation. The different forms of chronology are taken as elements in the process of turning a number of facts into a coherent historical narrative. In what follows, I will attempt to reflect on the ideological underpinnings of the different uses of time that appear in Japanese accounts of the past. In the final section, I will offer a few reflections on German historical scholarship in order to put what I will call the 'chronometrical turn' of Japanese historiography after 1945 into a broader comparative perspective.

The Poetics of Chronology

The periodisation of historical time is one of the central components of historiography. Different periods with clearly discernible features are distinguished by historical breaks and turning points. Thus, the flow of events is broken down into separate phases that, nonetheless, allow for the concept of overarching continuity. No periodisation, however, can do completely without any interpretive bias. In historiographical practice, the temporal co-ordinates are not a neutral framework but incorporate, rather, all kinds of assumptions and presuppositions that pre-structure historical interpretation. Periodisation of the past, thus, not only divides the wealth of historical material into manageable pieces; it also underlines the autonomy of different epochs as well as the possibility of long-term continuity.

The periodisation of history is intended to reflect the changes and caesurae observed in the evolution of the past. At the same time, it supplies an interpretive context that tends to invest individual occurrences with meaning even before they are actually interpreted. Subdivisions like 'the post-war period' or 'the Ancien régime,' for example, suggest an historical background in which events and developments seem to have obvious connotations. The same holds true for terms like 'the nineteenth century' or 'fin de siècle.' On the surface, they simply reflect the rules of the Gregorian calendar, while in fact, they are indicative of a certain *Zeitgeist*. Especially the latter concept points to the fact that periodisations not only reproduce the patterns of historical reality, but sometimes even shape them.

Another example of the poetic power of periodisation is the general division into ancient, medieval, and modern history that has become canonical in academic history. This pattern of organising the past has introduced elements of cyclical time into the basically linear concepts of time prevalent in the European tradition. Antiquity usually appeared as the 'Golden Age,' characterised by complex political institutions and a thriving culture. In many countries, antiquity had to be 'co-opted' since no ancient civilisation had existed on the territory of the modern nation-state. Thus, for want of alternatives in most Western European countries, Greek and Roman culture became part of the national tradition. In stark contrast, the Middle Ages represented a phase of decline. Medieval society, within this admittedly very general pattern, was usually characterised by inner conflict and lack of unity. Against this background, the age of Renaissance represented itself as a return to the origins, as an organic continuation of antiquity. The modern period, finally, made a resurrection of national traditions possible, finding fulfilment in the nation-state of the nineteenth century. The cultural traditions of antiquity were taken up by modern societies after a period of medieval decay.[1] This myth of a grand historical cycle connecting the bright ancient civilisations, via the detour of the dark Middle Ages, to enlightened modernity (a way paved by Renaissances and revolutions) also supplied the grand narrative of modern historiography.

Accordingly, the departmental organisation of historical scholarship at European universities reflected this general pattern. History departments were subdivided into units of ancient, medieval, and modern history. In the late nineteenth century, this structure was adopted by Japanese universities as well. In the 1880s, a department of history was set up at Tokyo Imperial University, and the Japanese government commissioned a young German scholar to institutionalise the discipline according to European precedence. Consequently, Japanese history (kokushi) was differentiated into antiquity (jôdai or kodai), the Middle Ages (chûsei), and the modern period (kindai).[2] Apart from organising historical scholarship, this framework also contributed to a sense of historical unity of the Japanese nation. The pattern of ancient, medieval, and modern times alluded to the myth of the succession of original purity and perfection, of a phase of national schism, and of the awakening to national subjectivity. Within this framework, it was possible to describe Japanese history as a process of national renaissance; the modern nation-state founded in 1869 was praised for overcoming internal strife and rebellion and for resurrecting the central authority of the state against local potentates. In a similar vein, the restitution of Imperial power that had been usurped by the Shogun for several centuries was seen as a return to the cultural origins of the Japanese nation. In yet another version, the modern period was interpreted as the rejection of the Chinese cultural and political hegemony over Japan, which had started in the sixth century with the import of Buddhism and the Chinese script. The military victory over China in 1895, thus, marked the possibility to revive the autochthonous traditions of the Japanese people. In all of these narratives of Japanese history, the general scheme of periodisation—the succession of ancient, medieval, and modern times—served to reinforce a sense of national unity over time.

The debates about the idea of the nation (minzoku) that flourished in the early 1950s may be taken as an example for this modelling of Japanese history according to the myth of a return to the origins. The outbreak of the Korean War caused apprehensions and warnings against a renewed American imperialism in East Asia. In this context, the idea of the nation as a site of resistance acquired a new significance, even among the majority of Marxist historians. In the annual meetings of the Marxist historians' association (Rekishigaku kenkyûkai), this prob-

---

1 See Duara 1995.
2 In historiographical practice, however, dealing with the modern period was largely eschewed by Japanese historians. Most universities, therefore, substituted the modern period with a chair for early modern history (kinsei).

lem of consciousness was expressed by focusing on 'The Problem of the Nation in History' (rekishi ni okeru minzoku no mondai, 1951) and 'The Culture of a Nation' (minzoku no bunka, 1952). One of the main features of these discussions was the search for a national essence that appeared to guarantee the continuity of Japanese ethnicity and culture. The historian Toma Seita, for example, in his study entitled 'The Origins of the Japanese Nation,' described a homogenous Japanese people that he characterised as inherently peaceful, dedicated to gradual development, and averse to military adventures. According to Toma, it was due to the import of Chinese elite culture that Japan had departed from its indigenous tradition and entered a fateful path that eventually culminated in the Second World War. He called for a return to the great cultural achievements and national values (most notably the originally pacifist disposition of the Japanese people) that had been lost during Japan's 'long Middle Ages.'[3] Likewise, the discussions initiated by the historian Ishimoda Shô about an ancient 'heroic era' (eiyû jidai)—presumably a golden age of indigenous folk culture that formed the unchanging core of the Japanese nation—adhered to the same overall pattern of periodisation.[4]

The Demarcation of a National Space

This general plot of antiquity, the Middle Ages, and modernity was introduced into Japanese historiography in the process of the institutionalisation of academic history. Within this (imported) framework, however, Japanese history was structured according to a specifically Japanese mode of organising time. For periodising the past, Japanese historians had recourse to the dynastic order of the Japanese monarchy. The reference to dynastic change was common practice in Chinese history, and the Japanese pattern of dating was clearly shaped by the Chinese model. Unlike in China, however, the Japanese monarchy ruled the country in an allegedly unbroken line going back to the mythical year of 660 BC. As a result of this continuity and for want of internal caesurae, the history of the monarchy did not lend itself to periodisation. Instead, the changing location of Imperial, and in later years Shogunal, power was referred to when speaking of historical periods. Depending on the place of the 'capital,' Japanese history was divided into the Nara, Heian (Kyôto), Kamakura, and Edo (Tokyo) periods.[5]

Thus, while the periodisation of the past implicitly alluded to the actual seat of political power, the dating of particular historical events according to era names (nengô) always referred to the current emperor. This negligence of the de

---

3 Toma 1951. See also the discussions in Rekishigaku 1951; also Duara 1998.
4 Tôyama 1968, 99-105.
5 At times, also the Shogunal dynasty supplied the name that demarcated a historical epoch, as in the case of the Tokugawa period (= Edo period, 1600-1868).

facto ruler and the emphasis on the imperial line marked a clear departure from the Chinese model. The era name was promulgated by the *tennô* (or 'emperor') upon his accession to the throne, and it labelled the years of his reign. The Shôwa (enlightened peace) era, for example, was inaugurated by Emperor Hirohito and denoted the period of his reign from 1926 until 1989. All events of modern Japanese history, according to this traditional pattern, were dated with reference to the Meiji (1868-1912), Taishô (1912-1926), and Shôwa (1926-1989) periods.[6] Without recourse to the regularities of the Gregorian calendar, therefore, the dating of historical occurrences followed an entirely internal logic. The periodisation and dating of the past was tied to the dynastic development of the Japanese monarchy and thus created a closed discursive space of Japanese history. Apart from organising the historical material, labels like 'the Meiji period' served an interpretive, or rather ideological, purpose. As the principal reference point of the temporal order of history, the figure of the Meiji Emperor seemed to infuse the years of his reign with a specific character of its own, with a certain *Zeitgeist*. This synthesising feature of the era names made it possible to refer to Meiji art, Meiji time mentality, or Meiji politics as unifying concepts; the term 'Meiji literature,' for example, seemed to imply that a variety of literary works shared a number of distinct qualities. The mode of periodising and dating suggested a cultural homogeneity in all spheres of society and created the image of a coherent and self-contained phase of Japanese history.

This 'traditional' pattern of dating by reference to the dynastic succession adhered to the Japanese custom of periodisation before the import of a modern, European-style historiography. It should not be seen, however, as a mere relic of a pre-modern preoccupation with the ruler in a feudal society. This pattern also corresponded to the historicist belief in the singularity and specificity of historical epochs. In the age of a 'modern' historiography, the traditional periodisation with respect to the reign of the *tennô* was compatible with the Rankean concept of the 'spirit of an age.' This brings us to a second aspect of this Japanese order of historical time. Not only did it serve to create the idea of homogenous historical periods, but it was also instrumental in what I would call the 'demarcation of a national space.' The specific chronology of history was conducive to a specific image of 'Japan,' the image of a nation that developed within its geographical bounds, independent of external interference. Japan, thus, appeared as a closed and unified entity whose history unfolded organically and autonomously. The turning points and caesurae of Japanese history, referring to the internal

---

6 The unity of Imperial reign and the respective name of an era, however, is a phenomenon peculiar to modern history. Before 1868, era names could change several times during the reign of the same emperor. These changes of political maxim expressed in an era name were an instrument of political propaganda and were used to announce a new phase of social reforms, or the end of a period of crisis or rebellion. The 21-year reign of Emperor Godaigo (1318-1339), for example, was characterized by seven changes of era name.

succession of the Japanese monarchy, seemed to lack any connection to events outside the Japanese archipelago. The traditional mode of periodisation, thus, contributed to the identity of a geographically coherent Japanese nation.[7]

The Quest for Universal Time

After World War II, Japanese historiography underwent a drastic change, and in the process, a new form of periodisation became canonical in historical studies. In the wake of military surrender, historical materialism soon emerged as the dominant influence on the interpretation of Japanese history. During the war, Marxist thought had been suppressed and most of its representatives were banned from the academy. After 1945, the tide had turned and Marxism now appeared as the only intellectual current that had not been implicated in the nationalist propaganda of the war years. For this reason, the American occupation army co-operated readily with Marxist intellectuals. At least until 1948, the occupation officers treated the Marxist opposition as welcome allies against the fascist elements in Japanese society. The Japanese Communist Party, in turn, greeted the American forces as liberators from wartime oppression. The social reforms initiated by the American occupation were hailed by some Japanese Marxist historians as the moment of a bourgeois revolution in Japan. In these first years after the war, Japanese politics witnessed an unlikely coalition of American occupation politics and Marxist intellectual discourse. At the universities, this resulted in the relegation of the most ardent representatives of a nationalist historiography. In their stead, Marxist economists and historians soon were called back to the institutions that had relegated them a decade before; within a few years, their interpretation of modern Japanese history was well established as the leading paradigm in Japanese historiography.[8]

This post-war emergence of a Marxist hegemony in historical studies can be seen as a major attempt to supersede nationalist interpretations by an international and potentially universal perspective. Before 1945, ultra-nationalist approaches had played a major role in Japanese historiography. The so-called *kôkoku shikan* historians, such as Hiraizumi Kiyoshi, had presented a decidedly Japan-centred picture of the past that rested on the nationalist *kokutai* ideology. The allegedly unbroken line of succession of the Imperial dynasty, unparalleled in the world, was the cornerstone of a version of the past that aimed to establish the superiority of the Japanese nation. During the war, these interpretations

---

7 See Karatani 1993.
8 This coalition of American occupation politics and Japanese Marxist historiography did not, however, last very long. As a result of the so-called 'reverse course' in occupation policies in 1948, the raison d'être of virtually every program of the Occupation was its contribution to the struggle against communism in and outside of Japan. See Schonberger 1989.

served to legitimate Japanese military hegemony in East Asia.[9] With the Japanese defeat in 1945, however, these attempts to enshrine the 'Greater East Asia Co-Prosperity Sphere' as a legitimate product of history came to an abrupt end. As a result of Japan's unconditional surrender, the nationalist ideology lost most of its power. The political purges during the American occupation, moreover, led to the removal of the most prominent proponents of kôkoku shikan from the universities and from public debate.

In post-war discourse, calls to transcend the narrowly nationalistic perspective of the preceding years were almost ubiquitous. Marxist historians were not alone in their demand to overcome the particularistic approach and to interpret Japanese history according to universal standards. After the nationalistic excesses of the war years, it seemed mandatory to put the Japanese past into a world historical perspective. Only six years after the war had ended, more than twenty new handbooks of Japanese history had already been published to establish a new, 'progressive' view of the national past.[10] Their quest for scientific objectivity, however, was not limited to the problem of adequate interpretation. The methodological tools of historical scholarship themselves were subjected to a thorough scrutiny. One of the central elements of this demand for universal methods was the issue of periodisation and dating of historical events.

The traditional periodisation of history with its implicit allusions to a closed national space was to be rid of its geographical aspects. Distinguishing epochs according to the seats of government ignored 'the commonalities of world history,' as Inoue Kiyoshi stressed in his popular 'Japanese History.' At the same time, it tended to overemphasise Japan's 'uniqueness' and to reinforce a 'pride in the incomparability of the Japanese state.'[11] For the purpose of integrating Japan into an all-encompassing world history, it seemed, above all, necessary to adopt a universal periodisation. This temporal order of history was not, however, a passe-partout that divided all histories into identical phases; the general periodisation had to be adapted to the socio-economic realities of each nation. From a world historical perspective, for example, the Russian Revolution in the year 1917 marked the beginning of the 'contemporary period,' while in Japan, the social reforms after 1945 seemed to be the equivalent historical turning point. The periods of world history, and consequently also of Japanese history, followed the well-known Marxist pattern: from primitive society (genshi) to the ancient slave-holder society (doreisei), to feudalism (hôken shakai), and to capitalism (shihon shugi shakai), with various intermediate stages. According to Ienaga, the 'positivist' (jisshô shugiteki) reference to the accidental facts of political history

---

9 For kôkoku shikan historiography, see Nagahara 1983, Bitô 1984, and Saitô 1984: 87-110.
10 See Shigakkai 1951a; apart from these surveys of general history, numerous handbooks of constitutional history, social history, and cultural history are listed. See also Yamada 1996.
11 Inoue 1963: 7-8.

were replaced by the universally applicable epochs of world history. In post-war Japan, this mode of periodisation with reference to the supposed social and economic realities was soon established as the standard procedure. Even non-Marxist historians would find it increasingly difficult to neglect this new paradigm; the conservative historians' organisation, Shigakkai, for example, adapted, in 1951, its basic textbook of Japanese history to the Marxist stages of development.[12]

Thus, the world historical perspective called for by Marxist historians consisted not so much of a geographical expansion of the analytic scope, but rather of a restructuring of chronology. The global character of history seemed to imply that all events could be mapped with reference to a universally applicable time frame. The pattern of periodisation was, in a sense, not the result, but rather the precondition of all interpretation. The temporal order of the past was, thus, not so much an instrument of description, but had an important poetic function as well: the general scheme of periodisation supplied the phases, the stages, and the turning points of history, and the actual events were then inscribed into this system of temporal co-ordinates. The stages of world history appeared as a given, and interpretation consisted in relating historical facts to this overarching structure.[13]

The problematic, thus, of assigning an adequate periodisation to the mass of inchoate events came to occupy centre stage in Japanese historiography. The majority of academic controversies of the early post-war years can, in fact, be reduced to questions of periodisation.[14] Debates about the beginning and end of feudalism, the period of manufacturing, or the exact dating of Japan's entry into the age of imperialism abounded in the 1940s and 1950s. The most important of these discussions was the 'Debate on Japanese Capitalism' (*Nihon shihon shugi ronsô*) which essentially focused on the historical status of the Meiji restoration. This controversy had originated among Marxist historians in the early 1930s; for a number of years, it was interrupted as a result of ultra-nationalist suppression, only to be picked up again after the war.[15] For the pre-eminent Marxist historian Tôyama Shigeki, it was of greatest importance:

> ... whether the Meiji restoration should be considered a bourgeois revolution. If we assume it to be a bourgeois revolution, the Meiji restoration marks the transition from feudalism to the capitalist formation of society; from this perspective it is evident that the Meiji restoration denotes the beginning of modernity. But if we refute the thesis of the bourgeois revolution, and if we

---

12 Shigakkai (ed.) 1951b.
13 For an interesting analysis of this mechanism, see Laclau and Mouffe 1985.
14 See, for example, Nagahara 1971; Tôyama and Nagahara 1963.
15 The best account in English is Hoston 1986.

consequently assume that the kind of power structure set up by the Meiji restoration was absolutism: then the issue of when the modern period set in becomes a highly difficult one, since absolutism is usually taken as the last stage of the feudal state.[16]

The rônô faction among Marxist historians treated the 1868 coup d'état as a true bourgeois revolution that ushered in the beginning of the modern period. A coalition of lower samurai and the upper strata of the peasantry, from this perspective, were responsible for the overturning of the old regime and for a series of social reforms that turned Japan into a bourgeois society. The rival kôza faction, on the other hand, stressed the feudal remnants of Japanese society that persisted until 1945. The promises of the Meiji reforms, in other words, were never fulfilled: the state structure remained absolutist, and the land reform did not liberate the peasants, but tied them even closer to a small number of absent landlords. It was not until the social reforms under American occupation, according to kôzaha historians, that these feudal elements in the social fabric were finally eliminated. In all these controversies, however, the underlying periodisation was not subject to debate. The succession of historical stages, in other words, was taken for granted, and the historian's task consisted merely of relating the empirical facts to the given temporal order.

Thus, the paradigmatic shift in the concept of periodisation palpably affected the practice of historical scholarship. Moreover, in order to supplement the 'universal' mode of periodisation, the dating of individual events was soon brought up to international standards. In a Marxist handbook published in 1949, the question of dating was explicitly mentioned: 'Because the contemporary world has become one, it is inconvenient for our country alone to be using a different chronology. Also, a thorough understanding of the past will prove difficult without a knowledge about what happened in the world at that time. Therefore the epochs of world history must from now on be unified.'[17] Therefore, it seemed obligatory to discard the Japanese practice of referring to the reigns of emperors. Instead, adopting the Western calendar was suggested as the proper approach to ensuring methodological rigidity. To make sure the reader understood, the handbook went on to explain how this new system worked: 'A hundred years add up to a century. Today we live in the twentieth century [...] and the fifth century before Christ was 2500 years ago.'[18] The introduction of the Gregorian calendar promised to set Japanese history into a world historical perspective.

---

16 Tôyama and Nagahara 1963: 210.
17 Minshu and Rekishigaku 1949: 5. Authors include such well-known historians as Ishimoda Shô, Izu Kimio, Inoue Kiyoshi, Eguchi Bokurô, Toma Seita, Tôyama Shigeki, Nagahara Keiji, Hayashi Motoi and many others.
18 Minshu and Rekishigaku 1949: 2.

Obstacles to Transition

In post-war Japan, thus, the traditional order of historical time was gradually replaced by a new mode of periodising and of dating the national past. This shift should not, however, be seen as a succession of two distinct 'periods' in Japanese historiography. Rendering the dates according to the Western calendar was not entirely uncommon in the pre-war years, and using concepts like 'the Meiji period' or dating with reference to the reigns of emperors is still common practice among many Japanese historians. August 15, 1945, the day of military surrender, did not mark a clear caesura in historiographical practice. Although in Marxist discourse the end of the war was often referred to as a 'Zero Hour,' as the moment that separated past and present, continuities in historical discourse nevertheless persisted. All this said, it is still striking that in the post-war period, dating with reference to dynastic development became increasingly rare. The pre-war years of the Shôwa period (1926-1989), for example, are often identified with the 'Shôwa 10s' or the 'Shôwa 20s;' referring to the 1970s as the 'Shôwa 40s,' however, is entirely uncommon.[19] This gradual change is not to be confused, to be sure, with 'progress' from a pre-modern to a modern pattern of periodisation. Instead, the shift from a 'parochial' to a 'universal' mode of ordering historical time should be understood as a shift in emphasis, a shift from stressing the uniquely Japanese character of the national past to stressing its international, world historical aspects.

The transition from a dynastic to a world historical mode of periodisation, which was a political concern for many 'progressive' historians after World War II, did not always go as smoothly as intended. A good example of the problems this transition produced is the concept of a contemporary history (*gendaishi*) in post-war Japan.[20] In a sense, this new term was meant to replace the most recent epoch in the standard dynastic pattern of periodisation. According to traditional practice, the year 1926, the year of Emperor Shôwa's accession to the throne, marked a new period in Japanese history. The post-war Marxist historians, however, set out to put an end to this form of reverence to the ruling class. In their eyes, dynastic change possessed no significance for the periodisation of history. This is clearly expressed in the book *Shôwashi*, one of the inaugurating works of post-war Japanese contemporary history. This controversial volume by Tôyama Shigeki, Fujiwara Akira, and Imai Seiichi, published in 1955, initiated a heated debate on how to write Japan's recent history. In 1959, the three authors felt compelled to react to the raging criticism and published a revised version of their

---

19 See Karatani 1993: 292.
20 At first, the term *dôjidaishi* was used as well; soon, however, *gendaishi* (or 'contemporary history') emerged as the standard terminology.

book.²¹ Given this broad public response, the pronounced departure from traditional modes of periodisation was not merely a moot point. The internal caesurae in the history of the Japanese monarchy, the authors claimed, were entirely irrelevant for both the development of society and for distinguishing historical phases. In a language that was reminiscent of the symbolic dethronement of 'Citoyen Capet,' they pointed out: 'In modern states, the death of an individual emperor (tennô kojin) is no more than an accidental occurrence, without any significant relations to the historical processes, and therefore without any relevance for the periodisation of history.'²² Significant events, in this context, were the outbreak of World War I, the Russian Revolution, and the so-called rice rebellion (kome sôdô) (massive protests against the famine raging after the end of the war in the summer of 1918).

This departure from traditional modes of periodisation, however, was not complete. Between the lines, the co-ordinates of the received chronology still exerted their power in the shaping of historical narratives. Tôyama, Fujiwara, and Imai deliberately employed the dates of the Western calendar, but in order to be sufficiently understood, these dates had to be supplemented by reference to the traditional era names. Thus, even the beginning of the Shôwa era acquired a renewed historical meaning: although in Marxist theory the driving forces of history were located on a deeper level, even a surface phenomenon like the accession of a new monarch, ironically, did not seem completely irrelevant. When Tôyama, Fujiwara, and Imai discussed the possible caesura that marked the beginning of 'contemporary history' (gendaishi), they opted for the mid-1920s: a time when the inner antagonisms of Japanese capitalism became increasingly critical, and class conflict intensified. In this larger context, the concurrent accession to the throne of Emperor Shôwa in 1926, 'even though a purely accidental coincidence, nevertheless endowed this year with historical significance.'²³ The proclaimed departure from traditional patterns of periodisation was thus an ambivalent undertaking, with the traces of competing discourses shining through as on a palimpsest. It was, therefore, symptomatic that the most influential work of Marxist contemporary history, setting out to replace the obsolete pattern of dynastic periodisation, still relied on the rather traditional title of a 'History of the Shôwa Period' (Shôwashi).²⁴

---

21 Tôyama, Imai, and Fujiwara 1959. For an interpretation of the so-called Shôwashi controversy, see Horigome 1965.
22 Tôyama, Imai, and Fujiwara 1955: 3.
23 Tôyama, Imai, and Fujiwara 1955: 3-4.
24 These 'obstacles' seem typical for the transition from a traditional to a 'progressive' historiography. Other examples come readily to mind: Fernand Braudel, for example, chose as subtitle for his structuralist analysis of the Mediterranean the rather conventional qualification 'à l'époque de Philippe II.'

History as Chronometry

The supporters of a 'universal' periodisation according to the Marxist stages of world history presented it as a deliberate move away from a parochial and Japan-centred order of historical time. As a result of this reform, events were stripped of their 'Japanese' aspects and inscribed into a generalised temporal framework. Dating and periodising lost its regional connotations, and history was analysed solely in terms of time: in the practice of the historian, historiography turned into chronometry. Nations were referred to as 'forerunners' and as 'latecomers'; the evolutionary model of change made it possible to assign labels of 'modernisation' and 'backwardness' respectively. Differences between cultures and nations were readily transformed into 'progressive' or 'accelerated' development on the one hand, and 'stagnation' on the other. Moreover, 'remnants of earlier periods' indicated the confusion of chronology within the same society.

One of the instruments most frequently employed to explain differences in the presumably uniform process of modernisation was the topos of 'stagnation.' Picturing Asian societies as stagnant and immutable was an integral part of Western discourse about the Orient. In the works of Hegel, Marx, or Weber, the idea of an inherently backward Orient found prominent expression. China, for example, was essentially treated as an 'empire of duration' (Hegel), devoid of any internal development. Perceived differences between Asia and the West were thus explained by reference to the stagnant character of Asian history.[25] In Marxist theory, a specific stage of development was devised to describe these periods when historical time seemed to have come to a halt.[26] Especially in the 1930s and 1940s, Japanese Marxists frequently employed this topos to explain Japanese backwardness with respect to the West. Hani Gorô was among those who linked Japan's incomplete modernisation to the 'Asiatic mode of production' (*Ajiateki seisan yôshiki*), that is, a period of social immobility.[27]

More importantly, however, the concept of stagnation (*teitairon*) was used as a tool to account for the differences between Japanese history on the one hand, and Chinese or Korean histories on the other. China had long served as a model for Japanese culture and politics. After the Meiji restoration, however, the asymmetry in their mutual relationship was inverted: Japan had successfully embarked on the path of (Western) modernisation, while China remained a passive object of world politics. Japan's military victory over China in 1895 confirmed this hierarchy between the two nations. As a consequence, the topos of 'stagna-

---

25 See Turner 1974; 1978; Rowe 1985; Shiozawa 1970. See also Said 1991.
26 Among Japanese historians of Marxist bent, there was considerable debate over the question of whether the period of stagnation was an historical stage peculiar to Asian societies. See Goi 1976 (especially chapter 3.3.), Fogel 1995, and Hoston 1986: 127-178.
27 See Naruse 1977: 43.

tion' was increasingly used to characterise Chinese, instead of Japanese, history. Japanese wartime historiography was particularly receptive to these prejudices elevated to the status of theory, but the idea of a stagnant Chinese society dominated post-war interpretations of Chinese history as well.[28]

While Chinese history thus appeared to be characterised by non-movement, Japanese history was increasingly depicted as the incarnation of speed and acceleration. This, to be sure, had not always been the case. The Marxist historian Inoue Kiyoshi reminded his readers that in pre-historical and ancient times, Japan had distinctively lagged behind the great civilisations of the earth. Compared to Mesopotamia, ancient Egypt, India, or China, the transition from primitive culture to civilisation in Japan was behind by close to four millennia. But these days of backwardness were over: 'Once this stage of development had been reached, Japanese society—even though not free of periods of precipitated development on the one hand, and of stagnation on the other—made rapid progress.'[29]

This reference to periods with different speeds of development brings us to a central argument in Marxist historiography. Societies were typically depicted as developing according to a predetermined set of historical stages; with respect to this allegedly objective set of co-ordinates, development could be rapid or slow, progressive or backward. Not all sectors of a given society, however, always flourished or declined at the same time. This problematic seemed especially conspicuous in Japanese history. Rather than developing in a coherent and homogenous fashion, Japanese society seemed riven with unequal development. Swift modernisation in the economic sphere, for example, corresponded with, in Marxist terminology, 'feudal remnants' in the political arena. The speed of development, then, was not the only parameter with which to gauge the success of modernisation. Instead, the notion of an appropriate speed, the ideal of sound and even development, permeated the writings of most Marxist historians in early post-war Japan. There appeared to be, as it were, a speed limit for successful modernisation, as is evident in Ienaga Saburô's observation that 'there was something unnatural about the rapid progress Japan achieved in a comparatively short space of time.' The imbalanced character of Japan's modern history, according to Ienaga, had resulted in internal 'contradictions' and structural 'defects.'[30] The heterogeneity of multiple speeds had turned Japan's modernisation into a distorted modernity (yuganda kindai).

This interpretation was widely influential in Japanese historiography for well over a decade after the war. In the late 1950s, however, the situation gradually

---

28 See Yoon 1995, 1989; Kan 1995, 1996; Saitô 1984: 147-172; Ubukata, Tôyama, and Tanaka (eds.) 1966.
29 Inoue 1963: 1.
30 Ienaga 1953: 223.

began to change. Japan had concluded a peace treaty with the United States and, thus, regained its sovereignty; moreover, the buds of economic growth were becoming visible, and the notion of continuous Japanese backwardness began to lose some of its plausibility. At this juncture, a number of revisionist scholars started to question the Marxist narrative of Japan's distorted modernity. Their reinterpretation of the national past rested on a different valuation of the speed of development and of the time schedule of modernisation. Kuwabara Takeo, a distinguished scholar of French literature who also specialised in the history of the French Revolution, most clearly expressed this change in perspective in his article entitled 'Tradition versus Modernisation' (dentô to kindaika) that was published in November 1957:

> The degree of modernisation can be assessed differently depending on which element is given greater weight, but here I want to focus on the speed of development of productivity. No other country before it had industrialised as quickly as Japan did, and a half-century later, only the USSR had come anywhere near achieving comparable speed of development. In considering the modernisation process as a whole, speed is one of the most important factors.[31]

In their endeavour to employ modern Japanese history as a success story, revisionists like Kuwabara met their Marxist opponents on their home ground. The paradigm of historiography as chronometry had become firmly entrenched in post-war Japanese discourse.

Chronometry in Comparative Perspective: The Elimination of Space

The 'chronometrical turn' we have described above was a prominent feature of historical studies in post-war Japan. It would be short-sighted, however, to treat this development as merely the historiographical dimension of an alleged Japanese peculiarity. On the contrary, similar developments can be traced in most parts of the world during the course of the twentieth century. The turn towards temporal explanations of difference as the privileged form of historical analysis may well be observed in a number of societies, Western and non-Western. We will shortly offer a few observations on post-war West German historiography to put the Japanese case in a broader comparative perspective. But before we do this, let us consider for a moment the impact of the 'chronometrical turn' on the relationship of space and time in historical discourse.

---

31 Kuwabara 1983: 40.

Interpreting the past above all on temporal terms not only introduced the idea of an appropriate and sound speed of development, but entailed certain unspoken consequences for the concept of space in historical analysis as well. As has been mentioned above, the traditional mode of periodisation, with reference to the succession of the Japanese monarchy, resulted in the demarcation of a national space within which Japanese history was seen to evolve. In stark contrast to this practice, the calls for a 'universal time' so frequent in post-war Japan implied, as it were, an elimination of space from historical discourse. According to the traditional pattern, historical dates had always implicitly alluded to the 'Japanese' character of past events; now they were transformed into neutral co-ordinates in a universal framework. The 'Genroku period,' for example, translated into 'the early eighteenth century,' and the 'Meiji 30s' was equivalent to 'the turn of the century' around 1900.

Thus, geographical reference was virtually eliminated as a relevant factor from historical discourse. The dates and periods that structured the historical narratives lost their regional specificity. Moreover, 'space' ceased to be the prevalent means for explaining differences in development. Deficiencies were no longer attributed to the mentalities of a people, and inferiority was redefined as backwardness. The cultural geography of the world was replaced by historical time as the sole factor accounting for the differences between nations. The concept of 'universal time' rested on the idea of a linear path of development that seemed open to all societies. Differences between them, consequently, could be explained by their relative position on that linear time axis. Where an older rhetoric had stressed the inherent and unchanging qualities of a people, their characteristics were now seen as the product of their 'progress' on the common path.

This kind of disappearance of space can be seen as concomitant to the 'chronometrical turn' in historiography. This was by no means simply a development of the post-war era. In fact, we can trace a genealogy of this mode of thinking all the way back to Enlightenment thought. The privileged status of time in historical analysis can be understood as the heir to the developmental schemes that entered historical discourse at the end of the eighteenth century. Montesquieu's comparative remarks on 'Oriental despotism' thus formed one point of departure for a range of social theories that were not content with describing difference, but rather situated people and nations on a time vector of development. The idea of universal progress was expressed by Turgot, among others, and was repeatedly reinforced in the course of the nineteenth century.[32] Social Darwinism only added to the ubiquity of this kind of analysis in the emerging social sciences. Also among historians—in, for example, the work of Henry Thomas Buckle—the developmental philosophy of history increasingly gained in importance. In Japan as well, historians like Fukuzawa Yukichi and

---

32 See, for example, Osterhammel 1998.

Taguchi Ukichi wrote 'enlightenment history' (*keimô shigaku*) that referred to a universal path of development.[33] The 'chronometrical turn' in modern historiography, thus, had a long history of its own. It was not, however, until the second half of the twentieth century and the emergence of the hegemonic position of social history that this mode of thinking advanced to the status of an unquestioned condition of historical discourse.

In this context, it is interesting to note that the emergence of social history did not always result in a systematic exclusion of spatial categories. The French school of Annales historiography, for example, was particularly known for its reference to geography as the most basic and deeply rooted structural condition of historical processes. Under the influence of Vidal de la Blache, the effects of weather and climate, mountains and plains, and the impact of rivers and the sea were introduced as explanatory categories. It was Fernand Braudel, in his book on the Mediterranean, who most systematically developed this train of thought and coined the term '*géohistoire*' to express the impact of geography on human history.[34] The ideas of Vidal de la Blache, by the way, not only influenced Lucien Febvre, one of the founding fathers of *Annales* historiography, but also had their repercussions among Japanese intellectuals. The cultural anthropologist Umesao Tadao, for example, drew explicitly on Vidal in his attempt to explain what he perceived as the striking parallels in the history of Western Europe and Japan. In Umesao's map of the world, these regions occupied the periphery of the Eurasian continent, a marginal geographical location he considered conducive to successful development.[35]

Thus, while '*géohistoire*' emerged as an important concept in significant parts of French historiography, the advent of social history in most countries was accompanied by a turn away from geography in favour of a concept of intrinsic development. This paradigm change can be observed in West German historiography as well. In the nineteenth and well into the twentieth century, the idea of history as determined by geographical conditions had been very prominent in German historical thought. German history was seen as deeply influenced by its location between Western and Eastern Europe, by its *Mittellage*. The topography of Europe seemed to make it more difficult for some nations than for others to develop organically, and the belated emergence of a modern nation-state in Germany was seen as but one consequence of these geographical conditions. Moreover, the German *Mittellage* seemed to imply a specific cultural identity, rejecting both Western European civilisation and Eastern European despotism. Likewise, the idea of an historical mission that Germany had to fulfil in Eastern Europe was deduced from geographical exigencies. During the Nazi period, this

---

33 See Ienaga 1954, Blacker 1954.
34 See, for example, Burke 1990.
35 Umesao 1957.

kind of geopolitical discourse exerted a strong influence on the expansionist character of German foreign policy.

As a consequence of their implication in Nazi politics, concepts referring to the determining impact of geography lost much of their plausibility after World War II. The substitution of time for space as the explanatory category (the 'chronometrical turn') was also a direct result of political changes in post-war Germany. To be sure: historians continued to write about German colonisation in Eastern Europe and the civilising mission that was supposedly its legitimising basis. The justification for intervention was still based on the notion of cultural difference between German culture and Polish and Russian barbarism. However, the reasoning behind this argument began to shift: the differences were no longer attributed to geographical location (and its concomitant, a stable and substantialist culture), but rather interpreted as the distance between 'peoples that in principle have experienced the same historical process of maturation, if only at different times.'[36] Thus, the temporal gap between national development—development that followed the same modernising path—seemed responsible for the differences between Germany and Eastern Europe. The cultural mission of the German people hinged on respective states of 'maturity' and no longer rested on cultural traits firmly entrenched in European geography.

In the course of the 1950s, this kind of temporal argument increasingly substituted an older rhetoric that had stressed the determinist influence of geography and climate. It was only in the 1960s, and with the gradual emergence of social history (Sozialgeschichte), that this 'chronometrical turn' came to constitute a new paradigm in West German historiography. Most social historians adopted the concept of modernisation in order to explain the structural similarities in development since the Industrial Revolution.[37] More importantly, however, this concept was appropriated in order to come to terms with the striking peculiarities of German modern history. Within the universal process of modernisation, the moment of departure, the early or late phase of economic 'take-off,' and the speed of development were perceived as the decisive parameters that determined success or failure. In order to explain the modern catastrophe of German history, therefore, reference was invariably made to Germany's status as a 'latecomer'—both in terms of industrialisation and the formation of a nation-state. The extensive debates on the peculiar German path into modernity (Sonderweg) were, thus, the most prominent manifestation of the prevalence of temporal arguments in West German historiography.[38]

---

36 Herzfeld 1962.
37 See Wehler 1975.
38 See Grebing, et al. 1986; Kocka 1988; Wehler 1981. It is interesting to note that much of the literature on the West-German 'Sonderweg' of the 1970s closely resembles the discussions of

Much like their Japanese colleagues, German historians after World War II turned to developmental theories of modernisation and to the concept of time as the privileged explanatory category. This fundamental shift that we have here termed the 'chronometrical turn' did not, unlike in Japan, translate into a different mode of periodisation. The end of the war in 1945 was considered the 'Zero Hour' (Stunde Null) by many contemporaries, but the formal frame of reference was not changed. Nevertheless, the shift of argumentative patterns was just as pronounced, and numerous basic tenets of German social history echoed the interpretive strategies of Japanese historians of Marxist and modernist bent. As such, when Reinhard Koselleck, who has contributed much to a growing theoretical reflexivity of German historians, introduced his terminological framework for historical analysis, it described the realities of Japanese discourse as well. Included were concepts like 'progress, decadence, acceleration or delay, the idea of not-yet and not-anymore, being comparatively early or late, being too early or too late, situations and duration.'[39]

Conclusion

After 1945, a new mode of ordering historical time became dominant in Japanese historiography. The traditional reference to dynastic change was gradually superseded by the 'universal' time of world history, the socio-economic stages of development, and the Western calendar. These modes of periodisation allegedly functioned as neutral systems of co-ordinates that were used to map the events of the past. As has become evident, however, they fulfilled an important task in the representation of the past as well. The different patterns of organising historical time not only served to register the historical facts, but also contributed to their coherent interpretation. Thus, the temporal order of things had strong poetic connotations and was a constitutive element of historical narration.

Different modes of representing historical time, therefore, are based on different sets of unspoken assumptions and ideological underpinnings. The shift in the mode of periodisation went along with a change in the ideological foundation of interpretation. In particular, we have observed a shift in the conceptualising of space that went along with the change in temporal order. The 'traditional' periodisation, according to the reigns of emperors, had contributed to a demarcation of national space. Dating and distinguishing periods by referring

---

Japanese Marxists in the early post-war years. The idea of feudal remnants, of 'late-comers,' and of structural deficiencies leading to a peculiar (and deviant) path of modernization when compared to the Anglo-Saxon West, was the leading paradigm in Japanese historiography a couple of decades before it became prevalent in West Germany.
39 Koselleck 1979: 133.

solely to internal caesurae created a closed national space within which all historical events seemed to originate. In the post-war period, many historians (often, but not only, of Marxist leaning) distanced themselves from this practice that they considered conducive to the nationalist ideology of the war years. Instead, they introduced what they called the 'universal' mode of periodisation in order to interpret the Japanese past as an integral part of world history. The dates and periods of Japanese development were stripped of all geographical connotations. What is more, 'space' was largely eliminated as a category of explanation. Based on a modernist philosophy of history that saw history evolve along a linear path of development, all differences between societies were attributed to a different temporal position with respect to this common path. This was by no means a phenomenon peculiar to Japan, however. In West German historiography, to give the example referred to above, we can find similar tendencies to replace geographical determinism by reference to time—as is true with most variations of modernist historical thinking.

These parallel developments, albeit strikingly similar, were not a simple reflection of what might look like a universal trend. Instead, the privileging of time as the explanatory category could take on meaning specific to the respective contexts. In post-war Japan, the 'chronometrical turn' was tied to a shift in the prevalent mode of periodisation. It was thus intended to replace a parochial, nationalist system of reference with a presumably neutral, universal framework. The 'universal time' of Marxist discourse, however, was a far from neutral instrument simply recording the events of the past. Instead, it was the result of a particular reading of European history that was then extended to the rest of the world. The succession of a slaveholder society, feudalism, and capitalism rested on an interpretation of the European past; clad in abstract terms, it was then presented as a disinterested pre-/de-scription of the non-European past as well. The 'universal' periods of world history, in other words, had a strong European bias. In a sense, then, the shift from 'traditional' to 'universal' time in Japanese historiography can be seen as a change from one set of references to another; the new mode of periodisation replaced the specificities of the Japanese polity with the specificities of Western development. It is no accident that this shift in emphasis occurred at a moment in world history when the attempt to establish a Japanese hegemony in East Asia had utterly failed, and the American occupation demonstrated the need to adapt to a global order again dominated by Western standards.

# Civilising the Past: Nation and Knowledge in Thai Historiography

Barend J. Terwiel

During the 1980s, I regularly attended annual meetings of the Thai Historical Society.[1] A large number of Thai historians gathered for these occasions, during which a general theme was discussed, invited speakers presented their papers, and matters concerning the Society itself were addressed. The events were quite formal: the speakers sat on a podium in a single row behind a long table, huge and apparently expensive bouquets of flowers decorated the free space near the podium, and gigantic banners hung on the walls, reminding all of the fact that they were there for the Society's annual general meeting. The speakers were usually well-known scholars, and the events themselves were important social happenings.

During one of these formal occasions, the large auditorium was filled almost to the last seat with those wanting to hear the three guest speakers. I estimated that between five and seven hundred people had come together. On this occasion, I had selected a seat close to the back, and I had the impression that I was the only *farang*[2] to attend the meeting that year. The panel sitting in front contained, as far as my recollection goes, apart from members of the executive committee, the speakers Professors Prasert na Nagara, Chetana Nagavajara, and Anan Ganjanapan. Prasert na Nagara opened the conference, Anan Ganjanapan spoke convincingly on methodological problems, and Chetana Nagavajara lived

---

[1] Its full name is Samakhom Prawatisat nai Phrarachupatham Somdetphrathepratanarachasuda Sayamboromarachakumari, or The Historical Society under the Royal Patronage of H.R.H. Princess Mahachakri Sirindhorn. Since 1978 the Society issues the journal *Ruambotkhwam Prawatisat* (Collected Essays in History).
[2] *Farang* is a Thai word with which all people of European descent are designated.

up to his reputation by presenting a witty and sparkling address that was much appreciated by the audience.

The chairwoman then allowed question time, and, from my position in the back of the huge room, I could only see the back of a person who stepped forward into the middle aisle asking to be heard. He had announced who he was before the microphone was carried to him so that his name remains unknown to me to this day, but I clearly remember his question which boomed crystal-clear through the hall with the help of sound magnification techniques. I here paraphrase the essence of his question:

> I am but a simple teacher of history, and I would like to be guided as to what I should do about the Nan Chao[3] Era of Thai history. I have heard that, among us historians, it has become widely known that the theory does not rest upon fact, that it has become commonplace to say that the Thais never governed the Kingdom of Nan Chao, and that consequently, it was not the conquest of Nan Chao which precipitated the movement of the Thais into Southeast Asia. The question of the origins of the Thais is believed to be quite different from that which is proclaimed in all textbooks. It is a fact, however, that all textbooks still proclaim the Thais to have passed through a Nan Chao Era. Can the learned members of the panel explain to me why the Nan Chao Era was not scrapped and why the textbooks have not been updated?

The chairwoman deftly passed the question on to Professor Prasert na Nagara, because, as she said, he was a member of the Prime Minister's Committee for History [Kammakan Haeng Chat Chamra Prawatisat, or the National Committee for Cleansing History].

Professor Prasert answered the question at length, and covered several points which I here summarise in the following manner:

1   It is true that the Nan Chao Theory has been attacked for some time. Some would even say that it has been discredited by historians who do research on this period.[4] Many would indeed say that the Thais had not been the rulers of Nan Chao.

---

3 Nan Chao refers to a political unit which had come into being during the first half of the eighth century AD and which lasted until its occupation by the armies of Kublai Khan in 1253. Nan Chao had its capital near Lake Tali in the present-day Chinese province of Yunnan, and modern scholarship reckons that Tai peoples never played a leading role in this kindom.
4 Among scholars it is now widely accepted that during the first millennium A.D., the Tai-speaking peoples lived in the coastal areas of what is now southern China. When the Chinese armies 'pacified' the region in the middle of the eleventh century, many Tai-speakers fled in a southern and southeastern direction to begin the conquest of many lowland regions of mainland Southeast Asia.

2   While this is the case, the first problem that arises is that scholars have not yet come to a new theory to which they unanimously adhere. At present, there are various conflicting hypotheses concerning the early history of the Thais. Some would even say that the Thais had always been here.

3   To rewrite the state textbooks, one needs a satisfying story. As long as that has not been accepted and formulated, we have kept to the old one.

4   One additional problem of a total revision of this aspect of the history books consists of the fact that when a new version would be decided upon, there would, for a period of time, be two incompatible stories, both with the seal of approval of the government. This would lead to confusion among school teachers as well as pupils: those using an older textbook would learn a different story from those who already possessed a new edition of the same history book.

The Birth of the Idea that Nan Chao Was Thai

The depiction of early Thai history as falling into eras, each connected with an empire-like political unit, goes back to the beginning of the twentieth century. At that time, many European historians liked to describe the history of mankind as being equivalent to a history of the world's constituent nations. In the first decades of the twentieth century, Thai historians, led by the chief minister Prince Damrong Rajanubhab,[5] accepted that model. Their task became to provide evidence that the Thai nation had a venerable past. As Crown Prince Vajiravudh remarked during his expedition of discovery in the northern regions of his country in 1908, he hoped that, through his report of ancient sites, 'the Thai will become more aware that our race is not a new race, is not a race of jungle folk, or to use the English word, "uncivilised".'[6]

This may help explain how eagerly the intellectual leaders of the time accepted information that confirmed a much longer history of the Thai nation than had been hitherto assumed. Particularly influential was W.C. Dodd's *The Tai Race* which appeared in 1923, with the proud subtitle: *Elder Brother of the Chinese*. Dodd extensively cites some ideas that were then current, such as that the ancestors of the Thai formed part of the Ugro-Altaic peoples that migrated several thousand years ago into the region now called China. Dodd describes for a period of more

---

5   Prince Damrong Rajanubhab (1862-1943) was a half-brother of King Chulalongkorn. At the age of thirty, he was appointed head of the Interior Ministry, a post he held until 1915. He played a major role in the setting up of a modern historical discipline by issuing historical documents and writing various books dealing with Siam's past.
6   Vajiravudh, Thiaw Mueang Phra Ruang, as cited by Vella 1978: 204.

than 2500 years no less than seven separate waves of migrations of the Ai-Lao peoples, whom he takes to be ancestors of the Thais. He assumes these precursors of the Thais to have been a great race well before 2200 BC, thus proudly assigning them a place among the oldest known civilisations of mankind. By AD 629, these Ai-Lao are described by Dodd as having developed the great state of Nan Chao, with its capital near Lake Tali in Western Yunnan. This great kingdom lasted until its overthrow by Kublai Khan, which Dodd (falsely) dated to have taken place in AD 1234.[7]

The foreword to the 1996 re-issue of Dodd's book stated:

> ... [the book] was eminently successful in propagating its extravagant scholarly claims. It provided Thai chauvinists of the late 1920s and early 1930s with an easily accessible set of information which revealed a glorious past, just when they felt in need of establishing such credentials for their state. For more than half a century Dodd's garbled early Thai history has dominated Thai-language sources that mention Thailand's early history. All of the bold claims in Dodd's book were adopted as part of the national history curriculum and generations of Thais have had to learn by rote how their civilisation was older than that of the Chinese, how they were to be identified with the Ai-Lao people in ancient Chinese chronicles and how they founded and dominated the Nan Chao empire, until Kublai Khan took away their freedom.
>
> Every one of these hypotheses was challenged by later scholars, but before such challenges could be properly debated, and a more balanced view be established, the whole flattering package had been worked into Thai textbooks and thereby gained an authoritative stamp of approval...[8]

It has now been generally accepted by historians that Nan Chao may, in its more expansionist phase, have incorporated some Tai-speaking minorities, but that it never was dominated by the ancestors of the Thais. Nevertheless, the myth of a Thai identity of Nan Chao had become part of the national curriculum and has remained so. It was still propagated in various Thai government publications during the 1970s and 1980s, and some prominent Thais have visited the region and were greatly moved to recognise traces of what they believed to have been their direct forebears.

In Professor Prasert's answer, he mentioned that some scholars claim that the Thais had always been there. At the time, indeed, there was a strong trend to propose this hypothesis, largely based on the fact that various indigenous peo-

---

7 For the correct date, see Footnote 3 above. Immediately following Dodd's book there appeared W.A.R. Wood's *History of Siam*, first written in 1924 and published in 1925. This book is silent on the earlier stages described by Dodd, concentrating instead on a glorification of the Thai 'Empire of Nan Chao.'
8 Terwiel 1996: xvi-xvii.

ples must have contributed both culturally and genetically to the mixture of peoples who at present share a Thai identity. This thought became popular only after 1969, after what was believed to be a breakthrough discovery on Thai soil that gave rise to a scenario that was even more spectacular than the 'elder brother of the Chinese' hypothesis.

Ironically, this even more spectacular scenario was, in its turn, also soon discredited by the scholarly world. This extravagant claim was that the very birth of human civilisation had actually taken place on Thai territory. This idea goes back to a single erroneous dating, published in 1969, in which it was claimed that a painted potsherd dating back to 4700 BC had been found at an archaeological digging in the village of Ban Chieng, and that iron objects from this period had also been found there. Since the manufacture of iron had hitherto been dated as much more recent, the claim—when true—would have had spectacular consequences for the whole archaeological world. This was indeed the theme of articles appearing in 1970 in *National Geographic* and *Scientific American*, as well as in newspapers around the world. These articles wrote of the need to rewrite the history of mankind, and to do so in such a way as to give credit to the people who once lived on the mainland of Southeast Asia for discovering the art of iron-making. However, further archaeological research did not confirm this extravagant claim: the oldest piece of painted pottery of the site now dates to 2860 BC, and all bronze and iron objects fall well into the bronze and iron age, thus relieving Ban Chieng from the burden of being the cause of a thorough rewrite of human history.

Undaunted by all disclaimers, a publication widely distributed by the Thai Prime Minister's Office in 1979 proudly announces:

> The world's oldest civilisation was flourishing in Thailand at least 5,600 years ago.
> Recent archaeological discoveries in the tiny, plateau hamlet of Ban Chieng, 500 kilometres north east of Bangkok, provide compelling evidence of a civilisation 600 years older than the ancient Tigris-Euphrates valley settlements, hitherto regarded as mankind's first 'Cradle of Culture.'[9]

Nevertheless, the short period during which it was believed that mankind's earliest iron-making had taken place in north-eastern Thailand had more effectively shaken the theory of the Thai's ancestry from the Altai mountains, and their subsequent presumed establishment of a Nan Chao empire, than decades of scholarly articles had done. A whole new room in the Bangkok National Museum was devoted to the new cradle of civilisation, and a growing proportion of publications no longer mentioned the idea that the ancestors of the Thais had

---

9 Thailand 1979: 9.

migrated from southern China, claiming instead that the Thais had always been living in Thailand.[10]

The Three Periods of Recent Thai History

While the earlier episodes of Thai history are now often couched in terms of one or both of the aforementioned, and rather flattering, hypotheses, the depiction of Thai history during the time from the twelfth century until the present has remained relatively constant since its model was constructed in the first three decades of the twentieth century. In one of the earliest overviews of Thai history, the period between 1257 and the present is divided into three eras (yuk). Largely inspired by the decipherment of the Rama Khamhaeng inscription,[11] the first period was found to be that when Sukhothai was the capital (1257-1350), followed by the Auytthaya Period (1350-1767), and finally the Bangkok Period (taken to begin in 1768 and lasting till the present). Each period in this series was named according to the city where the central authority was supposed to have resided at the time.[12] Moreover, it should be noted that the three periods are to be taken as subsequent, and that they are all three provided with beginning and ending dates.

This depiction of history is typical for the time: it presupposes the existence of complex centralised states and assumes, at the same time, a simple direct evolutionary line whereby the capital shifted gradually downward. Sukhothai was by far the most ideal state, under the wise guidance of a father-king. In order to organise its affairs in a satisfactory manner and in order to protect the state from dangerous enemies, Ayutthaya is described as needing a more absolute form of government. The Bangkok Period is the one during which paternalistic rulers successfully warded off all attempts by European nations to colonise the Thai nation, gradually educating the Thais and modernising them to the point where they could be trusted with a more democratic form of government. This division into three such eras and their perceived interrelation with particular royal pur-

---

10 See, for example, the article 'Khon Thai mai dai ma chak nai?' (Where Did the Thais Come From?) in Suchit (ed.) 1984.
11 The so-called Rama Khamhaeng inscription was discovered in 1833 by Prince Monkut, the future Rama IV (r. 1851-1869). At the time, it could not be satisfactorily deciphered, but in 1924, an authoritative reading was published by Georges Coedès. In the late 1980s, the hypothesis that the inscription was a fake (fabricated by Prince Monkut himself) was posed, and this caused a heated debate. In my opinion, the hypothesis must be firmly rejected, for the inscription contains information which was confirmed in other inscriptions which were subsequently discovered.
12 See Thi raluek BE 2468, 9. This Souvenir was a government publication that had been prepared for the international exhibition that was planned for 1925, but which had been cancelled when King Rama VI died.

suits has, in broad outline, remained unchanged and forms the framework of the history curriculum in primary and secondary schools.

In reality, as we shall see below, some primary sources were wrongly assigned, and the periodisation does not withstand historical scrutiny. To take the Sukhothai era as an example, the very beginning date of 1257 does not rest upon a firm basis. There is only one inscription that describes the conquest of Sukhothai by the father of Ram Khamhaeng, Pho Khun Ban Klang Haw, also known as Si Intharathit.[13] In this inscription, it remains unclear when this battle had taken place. Even if it appears that Si Intharathit replaced a Cambodian ruler, it is impossible to give more than an approximate date. The most accurate that one can be is to say that the city was acquired by the Thais around the middle of the thirteenth century.

It is also highly questionable whether Sukhothai should be depicted as a 'realm' or an 'empire' with a large territory of its own. This territory is shown in historical maps to have stretched from what is now northern Laos in the north to as far south as present-day Singapore, and from northern Vietnam in the east to Pegu in the west—a piece of land considerably larger than present-day Thailand. This imaginary realm was first created in the heads of historians during the early decades of the 20th century. It rests on a rather adventurous interpretation of a eulogy to Ram Khamhaeng that was probably inscribed not long after his death.[14] The list of towns which this particular king allegedly had conquered included Martaban in the west, and Nakhon Sithammarat in the south, but what manner of control he exercised and the extent of control over the regions lying between these cities needs to be considered rather carefully. It may also be of particular relevance that the town of Lopburi is not included in this list.

There is much to be said for the idea that, during the thirteenth century, there existed among the Tai-speakers many city states, some of them closely linked along trade routes, others forming temporary military alliances. This would help explain the sending of separate envoys by the rulers of Phetchaburi, Lopburi-Ayutthaya, and Sukhothai to the court of China.

Also highly problematic is the concluding date of the year 1350 which, in the government publication cited above, was taken as the end of the Sukhothai realm. This date was solely inspired by the fact that Thai sources place the beginning of the reign of Ayutthaya's King U Thong in the year 1351. Among Thai historians, it was quickly recognised that the Sukhothai era and the Ayutthaya era overlapped. As will be shown below, most historians accept that Sukhothai continued to exist for some time after 1351, but—it is here posed—as a result of the

---

13 See Griswold and Prasert 1972: 110-111.
14 There can be little doubt that the final part of Inscription 1 had been added some time after the main body of the text had been written. Epigraphists have deduced this from an abrupt change in the mode of address, but also by the fact that the final part is written in a different handwriting.

periodisation model, they still tend to minimise the period of overlap to less than fifty years.

The Myth of an Early Subjugation of Sukhothai

While the very idea of a Sukhothai Period preceding an Ayuytthaya Period has caused a bias to relate all that is written by and on Thais prior to 1351 to the history of Sukhothai, the opposite is also true: the concept of an Ayutthayan Empire, beginning in 1351, has caused historians to think in terms of a new centre, rapidly taking over from the previous centre. Thus, it is surmised that Ayutthaya 'soon' delegated Sukhothai into vassalage.

Inscription number 38, found in Sukhothai, written in Sukhothai-style characters, dated AD 1397, and promulgating a legal code of behaviour, has been taken to be the first proof of the ascension of power by the second king of Ayutthaya in this region. The name of the king in whose name this inscription was made is illegible; it has probably been deliberately erased. The scholars who have read and interpreted the inscription could only read the final two syllables, 'raja,' which is an epithet that is applicable to all kings of mainland Southeast Asia and beyond. The interpreters assume that the inscription was made by the ruler, who is referred to in the Luang Prasoet Chronicle as Phra Ram. At the same time, however, it needs to be noted that the Luang Prasoet Chronicle does not mention any military conquest for the reign of Phra Ram.

In the preamble of Inscription 38, it is written that the author of the inscription had recently conquered Sukhothai with an army which consisted of four sections drawn from four towns that were traditionally part of the Sukhothai realm. In their annotations, the translators comment: 'This passage is quite baffling. As the four places mentioned were part of the vassal kingdom of Sukhodaya [Sukhothai], their rulers presumably owed their allegiance primarily to the King of Sukhodaya and through him to the King of Ayudhya [Ayutthaya] …'[15]

Indeed, the idea that an Ayutthayan king would claim to have gained military control not through his own troops but by means of the troops that traditionally formed part of the core of Sukhothai armies is quite absurd. In addition, the choice of the words, 'vassal kingdom of Sukhodaya,' is only based upon the assumption that Sukhothai had already been subdued by Ayutthaya.

There are other translation problems that result simply from the presupposition that the inscription was written by the Ayutthayan king. For example, there is the fact that at the end of the preamble, the king expresses his desire to cleanse

---

15 Griswold and Prasert 1969: 129. In their article, 'Kingship and Society at Sukhodaya,' the authors express themselves a little differently and with more caution, saying that they read the mutilated name as the Ayutthayan ruler Ramarajathirat, but also noting that this identification is not conclusive. Griswold and Prasert 1975: 74.

the region in accordance with the ethical code in the manner of King Ramaraja. The translators are quite aware that in all other fourteenth-century Sukhothai inscriptions, the name Ramaraja is meant as the Sukhothai King Ram Khamhaeng. The translators naturally have problems explaining why this Sukhothai ruler is thus praised, and they assume that either the Ayutthayan king used this term to refer to the legendary hero of the Ramayana, or that the Ayutthayan king wanted to claim descent from Ram Khamhaeng in order to strengthen his claim on the region.

All problems to which have been alluded—namely the lack of a mention of a major victorious military campaign subjugating Sukhothai in the Ayutthayan Chronicles, the fact that the conquering armies consisted of troops drawn from towns immediately surrounding Sukhothai, as well as the puzzling reference to Ramaraja—are solved when we assume that the author of the inscription is a scion of the House of Sukhothai, a direct descendant of Ram Khamhaeng who described how he had eliminated a usurper and re-established order by proclaiming the law.

It would seem that the thinking in fixed eras and the blind acceptance of the model of successive empires have led Griswold and Prasert na Nagara into assuming an Ayutthayan identity of the author of Inscription 38. It is evident from the scanty sources at hand that there were many military campaigns in the region between Chiangmai and Ayutthaya during the fourteenth and much of the fifteenth centuries. Most, however, if not all of these may well have been raids for the sake of collecting elephants, manpower, and portable wealth, rather than as an expansion of a new 'empire's' territory.

The first firm indication of a territorial northern expansion on the part of Ayutthaya would appear to have taken place in the middle of the fifteenth century. This would have come with the decision of the Ayutthayan King Borommatrailok to create a second capital and for himself to rule the northern city of Phitsanulok, leaving Somdet Borommarat to rule in Ayutthaya.[16]

Ayutthaya Before 1351

Even though many older secondary sources still take as axiomatic that Ayutthaya was 'established' on March 4, 1351, it is becoming clear to a growing number of historians that it already was a thriving centre much earlier. The 'foundation date' of Ayutthaya in chronicles and inscriptions may, therefore, be better interpreted as marking an important political symbolic event, namely the sign of the establishment of a new reign, possibly symbolised by the establishment of a new palace. This would be in the same way as when, more than four hundred years

---

16 'Phrarachaphongsawadan' BE 2506: 136.

later, the first king of the Chakri dynasty moved his residence from Thonburi across the Chaophraya river. The legends surrounding Ayutthaya's 'founder,' U Thong, contain much information that supports the view that he was a usurper. Even his name is atypical for a Thai ruler. Most Thai history books have adopted the story that U Thong means 'golden cradle,' or 'a gilded cradle for the children of royalty.' It has been pointed out by Michael Vickery, however, that there are many other etymological possibilities and that the name remains rather enigmatic.

Another widely held and often repeated idea is that the term Xian (Hsien)—the word for Siam—in Chinese sources that deal with the region prior to the mid-fourteenth century refers specifically to Sukhothai. This was, for example, posed in 1904 by Pelliot, at a time when Thai historiography—under the influence of the then-current thinking of world history in terms of the history of nations—began developing the notion of a thirteenth-century Sukhothai Empire. From that time onward to the present day, Thai history has remained boxed into the aforementioned scheme that commences with a Sukhothai period and is followed by an Ayutthayan period (the latter, of course, conventionally taken to commence in 1351). Because Xian was the most important and most frequently mentioned Thai region in early Chinese sources, it was simply taken for granted that all references to Xian simply had to be references to the Sukhothai 'Empire.'

In 1958-59, Luce published a survey of the most important Chinese references to early Siam in which he translates the term Xian rather flexibly, sometimes as 'central Siam,' and other times as 'Siam, Sukhothai.' As such, he has shown that he was not altogether certain about equating Xian with Sukhothai.[17] E. Thadeus Flood, who has translated the early fourteenth-century *Jing-zhi da-dien* (or, *Ching-shih ta-tien*), may have been the first to express his uneasiness as to the identification of Xian as Sukhothai. However, by the very eminence and number of preceding scholars who have accepted that identification, he lets himself be persuaded of the notion that the identification with Sukhothai should remain.[18]

It was Vickery who first summarised the chief grounds for disallowing the previously generally accepted ideas about the identification of Xian as Sukhothai. He points out that the word Xian was evidently meant to indicate a region with one or more seaports in the Gulf of Thailand. Then, he notes that the very same Chinese sources that frequently mention their relations with people from Xian also mention not only Luo-hu (for Lopburi) and Bi-cha-bu-li (Phetchaburi) among the individual Thai cities, but Su-gu-di (Sukhothai) by name as well. After describing how an envoy from Sukhothai arrived at the Chinese court on June 15,

17 Luce 1958a; 1958b.
18 Flood 1969: 241, fn.82.

1299, they report in the very next entry (on July 7, 1300) the arrival of a group of people from Xian. Thus, it is unlikely that Xian and Sukhothai are identical.

In addition, it is notable in the entries of the Yuan Shi that the relationships between Xian and China are maintained by marine traffic, not overland, as would have been expected for Sukhothai. The ruler of Xian is described as carrying the title *gan-mu-ding*, which has been identified as the Cambodian title *kamrateng*, which may also be interpreted as a sign that this ruler came from a region that was directly influenced by Cambodia.

Vickery has quite convincingly argued that the very standard classification of the two succeeding empires, whereby Ayutthaya did not exist prior to 1350 in the conceptual schemes of many historians, has prevented these historians from making the much more logical identification of Xian with Ayutthaya. Vickery thus posits that the town may have been among the first occupied by Thais in the Chaophraya delta, and that it must have been a trading centre since the eleventh century.[19]

Yamamoto, who made a thorough study of the *Da-de Nan-hai zhi*, adds further evidence, not only to the view that Xian was Ayutthaya, but by pointing out that, at the beginning of the fourteenth century, Zhen Da-zhen considered Ayutthaya to be more important than Sukhothai.[20]

Corroborating evidence for Ayutthaya's early pre-eminence comes from the fifteenth canto of Prapañca's *Nagara-Kertagama* (1365), in which some of the countries surrounding Majapahit are listed. Siam is described with the following words: 'Syangkayodhyapura [my underlining], together with Dharmanagara, Maratona and Rajapura.' The most prominent position is thus accorded to Ayutthaya, an unlikely situation if the city had indeed been founded a mere fifteen years before.[21]

When accepting the thought that Ayutthaya-Lopburi formed an independent Thai principality prior to 1351, various hitherto-puzzling features in Thailand's early history can be reinterpreted in a much more satisfactory manner. For example, since Chinese sources indicate that a ruler of Xian had died about AD 1298, we need no longer assume that this refers to the death of Ram Khamhaeng, as Griswold and Prasert do, even when it brings them into conflict with at least two versions of the History of Martaban.[22] Also, if we are to accept an earlier identity of Ayutthaya as a Thai centre, it need no longer puzzle us as to why the oldest preserved Ayutthayan chronicle (the aforementioned Luang Prasert chronicle) commences with an entry that is dated CS 686, corresponding with AD 1324-1325, 26 years prior to the establishment of the reign of U Thong (Ramathibodi).

---

19 Vickery 1978: 204-205; 1979: 155 ff.
20 Yamamoto 1989.
21 Pigeaud 1960: vol 3, 17-18. See also Ferrand 1914: vol. 2, 663.
22 Griswold and Prasert 1972: 20.

If we accept this alternative identification of Xian as Ayutthaya, then the information from chapter 33 of the *Dao-yi zhi-lüe* stating that Xian attacked Lopburi in the fifth month of 1349 can be read as an indication of a period of political unrest and activity on the part of Ayutthaya, which may have culminated in the establishment of this new dynasty in 1351.[23] In addition, it need no longer surprise us that there is no mention in Sukhothai inscriptions of the attack on Cambodia which occurred during the thirteenth century.[24]

There has never been any doubt about the fact that, in later Chinese sources, Xian is Ayutthaya—such as in the chapter on Xian in the *Yuan Shi*,[25] as well as in Ma Huan's *Ying-ya sheng-lan*, which was written in the middle of the fifteenth century. Over time, Xian and Luo-hu (Lopburi) are recognised as having joined to become one and the same political unit, which gave rise to the name Xian-lo, the Chinese name for Siam.

Ma Huan explains that in travelling from Champa with a fair wind, the ship takes seven days and nights to reach the estuary that has been identified as the Gulf of Thailand, it then enters the anchorage, and one then reaches the capital. Ma Huan also mentions that the soil in Xian was barren, little of it suitable for cultivation, and that the area around it was wet and swampy.[26] Mills comments that, although this information is supported by Kung Chen, Ma Huan must almost certainly be wrong, for the central region is very fertile. However, in this case, I believe Mills has made an error of judgement; he has apparently been influenced by a mental picture of the central plain in modern times. During the fourteenth and fifteenth centuries, the lower Chaophraya region was most likely not yet developed for agriculture, as the canal system which opened the region post-dates this period.[27] When the word Xian is intended to mean the city of Ayutthaya, Ma Huan's description of the place as having insufficient rice fields corresponds with the Ming records, which state that Ayutthaya had to be supplied with rice from Lopburi in the north.[28] Altogether, then, it appears necessary to rewrite those passages in the standard textbooks that deal with fourteenth-century Ayutthaya to allow for an early Thai identity.

---

23 Rockhill 1915: 100.
24 Pelliot 1902: 51.
25 Grimm 1961.
26 Ma Huan, p. 103.
27 See Tanabe 1977. Tanabe tells us that there is hardly any mention of irrigation works in the earlier parts of the royal chronicles, and that during the early Ayutthaya Period, canal digging in and around Ayutthaya appears to be in the first place inspired by the need to improve transport and for military strategic reasons.
28 Grimm 1961: 2. See also Yamamoto 1989: 54.

The Authority of Knowledge

These two examples demonstrate how a particular model of thinking in the depiction of Thai history had been created more than fifty years ago, and how this model has survived without attracting a growing opposition. This model was created with the assistance of senior government officials, and it has been perpetuated in school textbooks and other publications by successive Thai governments (Figure 4). It has also been shown how adherence to this model of successive empires has contributed to a particular way of interpreting data that fits in with the model, and how it may have prevented alternative readings. We have shown that the presupposition that there was only one major Thai empire prior to 1351—that of Sukhothai—has led many scholars to assume that all Chinese sources referring to Hsien (Xian) dated earlier than 1351 simply had to refer to Sukhothai. Similarly, the thought that Ayutthaya took over as the region's new Thai empire not long after 1351 has caused leading historians to present one of the most important Sukhothai inscriptions as having been written by the king of Ayutthaya, despite the fact that this identification leads them to various translation dilemmas, and to regard a crucial passage as 'baffling.' Finally, the model has prevented historians from seeing the role that a pre-1351 Ayutthaya may have played.

The historical model that was developed shows the interesting tendency to idealise the past and to interpret the present as a somewhat tattered inheritance. The further back in time, the more signs of an ideal society have been noted. The Sukhothai period is invariably depicted as one during which the individuals enjoyed immense personal freedom. Any facts that run contrary, such as the fact that the society was definitely hierarchic and that there existed a class of slaves—facts that can clearly be drawn from the inscriptional record—are left out, while all idealised statements from the inscriptions gave rise to eulogies.

This idealising of the early past fits in with an indigenous attitude about the very sources of knowledge. The various branches of traditional knowledge (transmitted by teachers of various musical instruments, dance, masked theatre, and shadow theatre, as well as magical tattooing and traditional boxing) are seen as precious traditions that have been handed over by great teachers who used to live in an unspecified, deep past. For all these important branches of traditional knowledge, there exists a ceremony to evoke the old teachers. It is called *wai khru* (or 'to acknowledge the teachers'). It is a ritual that is usually performed as part of a final examination of a master class, during which the teacher kneels down, evokes the gods, and prays for their assistance and for guidance from those who were at the very beginning of the long chain through which knowledge has been transmitted. Implied in the ritual is that at some time deep in the past, the traditional skills were complete and perfect, and that the teachers at present are but intermediaries between this perfect tradition and the still-imperfect pupils.

Figure 4
One of a series of stamps—issued in 1983 to commemorate 700 years of the Thai alphabet—reproduces part of the beginning of the Rama Khamhaeng inscription.

This traditional concept of knowledge helps explain why so much emphasis is laid on rote learning in the traditional schooling, and why, in the vast majority of Thai schools, the teacher is considered to be unassailable in the role as transmitter. The teacher has been formally qualified to hand on parts of the wisdom of the ages, and in this task is guided by textbooks that have been checked and corrected by the nation's best minds. This traditional attitude towards knowledge does not leave much room for personal interpretation, nor does it foster a spirit of individual enquiry, let alone critical commentary. The attitude that teachers are transmitters of a well-nigh perfect body of knowledge is shored up by the very appearance of textbooks and editions of primary sources. In the first place, the textbooks are not presented as having been written by a particular individual, but are produced by anonymous government-appointed textbook committees. Also, the standard editions of historical primary sources, such as the major literary works of the past, are usually presented as a perfect composition;

seldom is there mention of the conflicting manuscript versions underlying a standard text, and even rarer is a footnote expressing doubt as to the standard reading. This traditional attitude toward the preservation and passing on of knowledge also helps explain why the teacher's question, paraphrased at the beginning of this chapter, was such an apt and pertinent one.

It is certainly not the intention of this article to depict the whole of Thai intellectual life as traditional, backward-looking, or stagnant. There are many Thais—especially those who have had the fortune of having been taught by teachers who emancipated themselves and who move freely in a post-renaissance world—who have internalised modern scientific methods. The teaching of a critical pursuit of knowledge, particularly in relation to the social sciences, was one of the principles behind the establishment of Thammasat University in 1934.[29] Thus, as demonstrated by the bold anonymous teacher's question mentioned at the beginning of this chapter, an appreciation of a critical spirit of enquiry certainly exists among intellectuals. Instead, I would like to draw attention to the fact that the many Thai intellectuals—those who have internalised the idea that human knowledge is fed by fundamental research, and that this knowledge is in a state of rapid expansion in unknown directions—often find themselves hemmed in by the consequences of a widespread conservative attitude as to what learning is about, and by the government's active role in using knowledge for its own nationalistic purposes.

Indeed, this article may draw attention to the plight of modern practitioners of the historical profession who work and conduct research in the intellectual climate of Thailand—where the Thai government has almost exclusively taken control of the national curriculum, and where the extent to which information on the past has been codified leaves little or no room for debate, let alone alternative historical models. If an historian in Thailand were to come across a document that stood in direct contradiction to the standard state-sponsored version of what happened, he or she would immediately be in a dilemma. On the one hand, it is apparent that, for history's sake, the document should be made public and its consequences debated in public. On the other hand, however, some powerful sections of the Thai academic world may frown upon the person who divulges such information, for it may be interpreted as disloyalty to the state. As a final comment, this situation may help explain why the chief challenges to the sequential empire model that has dominated Thai history for more than fifty years have not in the first place come from within Thailand, but chiefly from researchers living outside the borders of the country.

---

29 The founding of Thammasat University was the pet project of Pridi Phanomyong (luang Pradit Manutham) who, at the time, was a member of the cabinet. He planned Thammasat as a university for the people, as a counterweight to the older elitist Chulalongkorn University.

6 Splitting Historiographies in Aru (Indonesia)

Patricia Spyer

This chapter investigates three sites where Aruese formulate the splitting historiographies through which they negotiate their place at the frayed edges of modernity.[1] Each of these split sites is different in character and genealogy, and each poses different problems for interpretation. The status, demarcation, and relative values of 'past' and 'present' in each are crucial. The first example of a splitting historiography, and the one which receives the most attention here, deals with the problem of periodisation. Specifically, it deals with how a particular period, the so-called Dutch Era—or *Zaman Belanda*, as commemorated in song—stands as a signpost between a past of forceful Aruese leaders and powerful speech and a recent New Order filling the space where alternative voices were violently silenced and suppressed. The second example codifies the equally violent history of the coerced conversion of Aruese pagans to world religions in the mid-1970s, and offers a somewhat different reading of the institutionalisation of a break between the past and the present. While the first two examples suggest how the violence of historical experience, and specifically, that of the

---

[1] Twenty-four months of fieldwork in Aru (1984, 1986-88, 1994) were funded by a Department of Education Fulbright-Hays Dissertation Fellowship, the Wenner-Gren Foundation for Anthropological Research, the Institute for Intercultural Studies, the Southeast Asian Council for the Association of Asian Studies with funds from the Luce Foundation, the Netherlands Foundation for the Advancement of Tropical Research (WOTRO), and was conducted under the sponsorship of the Lembaga Ilmu Pengetahuan Indonesia and Universitas Pattimura. The Institute for Intercultural Studies also provided funding for three months of archival research in the Netherlands. I am grateful to these institutions for their generous support of my work. This paper was written for a conference on 'Time Matters' in November 1998 organised by Leo Douw, Willem van Schendel, and Henk Schulte Nordholt. I thank them and the other participants for their comments and suggestions. Rafael Sánchez offered valuable comments for revision and I thank him for this.

former Suharto regime, may become codified as a rift in historical memory, my third example considers how such a rift may be explicitly provoked in specific circumstances. I suggest how the theatrical staging of a 'coup,' in the context of an annual ritual performance, forms the prerequisite for the inauguration of this performance's 'other' time and space.

Aru is a low-lying archipelago made up of six large central islands and multiple smaller ones clustered under the Bird's Head of New Guinea in the southeastern Moluccas of Indonesia. For several centuries, the lives of women and men in these islands have been shaped within the erratic conditions of a volatile luxury product trade, extended networks of commerce and communication, and the political projects of successive colonial and post-colonial regimes. In the eastern parts of Aru, known as the archipelago's Backshore (as opposed to its western so-called Frontshore with the island capital, Dobo), relations to the economic and discursive forces of trade, the colonial *mission civilisatrice*, and the more recent Indonesian nation-state, have, in certain important respects, been incomplete and discontinuous—if at times also brutally invasive. It is this complex positioning—being simultaneously entangled within and marginalised to the nation-state—which accounts for the fragmented acquaintance Aruese have with nationalist historiography, on the one hand, and for their ability to bend it to some of their own purposes, on the other. In the same vein, this positioning explains why Aruese refigurings of such ready-made tokens of national historiography as the 'Dutch Era' inevitably betray the constrained circumstances of their production in one of Indonesia's many frontier spaces, while at the same time suggesting, once again, how people make their own history within conditions that are not of their own making.

The Signpost of Silence

Compared to other Aruese songs, '*Zaman Belanda*' is unique, both in terms of its idiosyncratic structure and in the distinctiveness which it assigns to the 'Dutch Era' it commemorates. Both idiosyncrasies are addressed in this chapter. The song documents the suppression of an important rebellion against colonial rule, the shattering of powerful words in this context, and the emergence of a song about the Dutch Era that, paradoxically, is the silent tomb of this event. I argue that the song marks an important depletion of Aruese authority, and the emergence of a silence at the moment of its own creation. I interpret this with reference to the relative suppression of discourse about the Dutch colonial period under the New Order, as well as with respect to the song's potential political implications—the possibility of insurrection against the powers-that-be intimated by the song, should potent words be allowed to circulate.

For many Barakai islanders in south-east Aru, the song is held to document an unusual, if relatively unelaborated, moment in Aru history—the *Zaman*

Belanda, or Dutch Era. What is most striking is that in contrast to the more official forms of nationalist historiography—which use periodisation to construct an ordered and clear chronology, and from which Barakai's *Zaman Belanda* is clearly derived—they recognise only a single period on Barakai. *Zaman Belanda*, in other words, stands entirely on its own, enjoying a unique position that is neither set off against, nor incorporated within the framework of a larger periodisation. While Barakai islanders clearly recognise the concept of a 'period,' or *zaman*, as an historiographical category, they apply this exclusively to the isolated and reified *Zaman Belanda*. In all other circumstances, any sense of periodisations or singling out of distinct eras does not apply. When women and men of Aru's Backshore speak, for instance, of the time of their own ancestors, they simply invoke it with some statement along the lines of 'long ago...' Similarly, stories and references to the Japanese occupation of Aru tend to be casually broached with remarks like 'when the Japanese were here...' I will return to this anomaly later; for the moment, I shall highlight only some of its peculiarities.

Handed to me on a white piece of paper with the words '*Zaman Belanda*' at the top and, as far as I know, never sung, the song already inscribed both the reification of a distinct Dutch era and its own silence in this manner of presentation. The song given to me by Lakulu Kobawon, a man from the Barakai community of Longgar and the person who wrote down the song and titled it *Zaman Belanda*, grants the particular phase of the Dutch Era the capacity to represent the whole. According to Lakulu, the song is about the capture by the Dutch of a member of his patriclan, or *fam*. He was a man named Pukulgore Kobawon, whom Lakulu claims was arrested in the 1890s as he returned to the village after a day of diving. It is, in fact, quite unusual for Aruese to refer to dates, but Lakulu, unlike any other Aruese I knew on Barakai at the time, had worked in construction in the provincial capital of Ambon and was, therefore, somewhat more attuned than other islanders to national discourses. During his stay in Ambon, he may have internalised some aspects of nationalist historiography, such as its chronologies. These chronologies are embedded within the organisation of public space throughout Indonesia, in the aesthetics and forms which the former New Order regime bestowed on its official monuments, as well as more explicitly in the curriculum of the national educational system.[2]

In referring to the 1890s as the time of a rebellion which might have involved one of his ancestors, Lakulu was not far off the mark. In 1893, a large-scale revolt involving Barakai islanders and many other Backshore peoples swept across the eastern and south-eastern pearldiving areas of the archipelago. The movement was only suppressed after Dutch battleships were brought in, several armed confrontations had taken place with some loss of life on the Aruese side, two villages had been reduced to ashes, and the rebellion's most important leaders had

---

2 Lindsey 1993.

been arrested by the colonial authorities. Beginning in the 1850s in north-east Aru, and then with mounting frequency between the early 1880s and 1907, the archipelago's Backshore peoples—who subsisted primarily on the seasonal collection of various 'splendid and trifling' luxury articles for trade such as mother-of-pearl shell, trepang, birds of paradise, and edible birds' nests[3]—rebelled periodically against what colonial sources tend to gloss over as 'Buginese and Makassarese' traders and the latter's 'conniving practices.'[4] While Dutch accounts mention that virtually all of the rebellions were directed against the colonial government in addition to the traders, they tend to highlight the causes of the disturbances as being either the exploitative practices that they associated with the Muslim traders, in particular the credit system through which the latter indebted Aruese to themselves, or the superstitious nature of the indigenous population.

In addition to whatever tensions existed between Aruese and the traders, the rebellions of the late nineteenth and early twentieth centuries clearly emerged in the context of the consolidation of Dutch rule on the islands (a local version of a much larger process of imperialisation that was taking place across the Netherlands East Indies at this time). Equally important, the movements also surfaced within a complex situation which seems to have involved the reshuffling of Backshore trade dynamics due to the novel pressures and competition that were introduced with the arrival of pearling companies from the neighbouring Banda Islands and Australia. These began to operate off of the Backshore in the 1880s and 90s.[5]

Of all the movements that swept the Backshore around the turn of the century, that of 1893 seems to have been both the largest and the most violent. The Colonial Report (Koloniaal Verslag) of 1893 contains a fairly detailed synopsis of the rebellion which began on the Backshore in September/October 1892, or at 'that time approximately when the Chinese and Makassarese merchants of Dobo and other Frontshore communities customarily come to trade.'[6] According to the report, twenty-three Chinese and Makassarese, including women and children, who had arrived on the Backshore at the beginning of the commercial season were killed, while others went missing. Although the report fails to specify the location of the killings, it is possible that they took place on Barakai Island itself, since a court case of June 1893 compensates twenty-seven traders representing all five communities on this island for damage to their houses and wares

---

3 Van Leur 1983.
4 Van Eijbergen 1866: 223, 229; Van Höevell 1890: 38.
5 Spyer 2000.
6 Koloniaal Verslag 1893: 25-28.

as a result of the rebellion.[7] On November 24, the day before the resident arrived in Aru, a fleet of eighty-five native *prahus*—each carrying fifteen to twenty men armed with krisses (probably machetes), bows and arrows, and spears—launched an attack on the Aru capital, Dobo. The attack was repelled by fifteen well-manned Makassarese vessels which had come to trade on the islands, as these were supported by the regents of several Christian communities near Dobo. It was, however, not until a good month later when two battle-ready steamships, the Java and the Arend, arrived at Aru, that the colonial government could put down the rebellion.

'*Zaman Belanda*,' the song, does not detail nor focus on the events comprising the rebellion, nor does it name the motivations thereof. Instead, the song focuses on the moment of suppression, and with the two being the same, the moment that the composition of the song was conceived. Along with the story of the song's own emergence, it tells of the arrest of an important leader being taken on board a ship, and of a meeting between the Dutch, the Buginese, and some Aruese headmen. Significantly, it locates the song's genesis directly in the context of pearldiving. Here is my translation of the Maluku Malay translation that Lakulu and I did of the Barakai song. I follow Lakulu's division of the song into four parts. The first goes as follows:

One:
Rasuk darnomnom kol-kol sien   A fleet of sailing boats dives off
                                                  of the white beach
dal tamenmen ru dam da eti   they drift on their anchors[8]
gwagwil malala Daldal   a young man dives pressed for time
Jamin asis em Logar sel   his father Jamin says to the son Daldal:
Logar selo   it will be decided in Longgar
dam sabe a gul   in the village of Longgar

---

7 Quite a number of the names are Chinese, while the title *Daen* (properly *Daeng*) before proper names indicates a Buginese origin. One name, Abdul Salam, is properly Arabic, while a few others like 'Honey,' *Madoe*, and 'Rat,' *Tikoes*, are presumably nicknames. According to the report, the traders due to receive compensation included nine residents in Apara (Abdul Salam, Langale, Batjo Kalankarie, Daen Pawara, Ingtai, Palalo, Manloeka, Moesa, Tamalaga), eight in Longgar (Tan Kwan Tjai, Oei Hie Hoan, Seng Tjoan, Daen Baoe, Madoe, Wa Kasso, Daoeda, Sempo), seven in Bemun (Daen Paliwan, Tonga Patola, Salewa, Tikoes, Palarie, Wanalisa, Batjo), one in Mesiang (Daen Mangesa), and two in Gomo-gomo (The Thie, Daen Matola) (Mailrapport 1893:6503, #858+, Algemeen Rijksarchief, the Hague).

8 Following Lakulu, this line and the following indicate that the divers are in a hurry since the news of the Dutch arrival in Longgar to arrest their headman, Pukulgore Kobawon, had already reached them. Instead of moving to another location and casting out their anchor again, they let their lines out and drift on their anchors.

asis wanu el rau	he began to compose in his head the song [about the affair] that would break the village's powerful speech

Two:
Gulne lirauo, lirnal	The chief speaks, he utters
Ambong rua dajo ngol-ngol	two Ambonese blow their whistles
gulne a ta	the chief [is led] to the sea
Gul daririn tan kader	the heads rest sitting in chairs
tongko Walad	[they have] Dutch colonial staffs of office
Bugis nal goggora	the Buginese are angry
Walad asis em mairun	the Dutch decide that today
kom guguli dajur Bugis	our heads and the Buginese
dam derwa rau	will exchange words

Three:
Sirgwa jamlir muwe mi kabal	Because of words spoken Sirgwa escorts him to the ship
Magur amam nam lelerwi	the father hushes Magar
kiolar malala wuwun	I climb up
kutmat kabal ken manara	I glance at the ship's bridge
tuliso ruba lei	the writing has changed
korkojamor kabal wun	I ascend to the top of the ship
kuliskol reson lau-lau	I climb up the ship's ladder
kawolko gulala murin	I go into the back of the cabin
kutmat rarar Ambong rua denden	I see two young Ambonese women sleeping

Four:
Ja murbano Lob koljurun	You go to the beach at Dobo
kabalio aulen lau	a ship is at sea
Ambong dua limnal gorgornean	the Ambonese head raises his eyeglass
eslal nal Logar dua	he scans and spots the Longgarese chief
evjin melmel	who is laughing a great laugh

Before commenting on the song, I want to say something about the man, Pukulgore, who was led to the ship to be taken to Ambon where he is said to have died in prison. I also want to say something about the particular fears he aroused, according to some of this man's descendants and a number of other islanders, in the Dutch colonisers. I first heard about Pukulgore from Lakulu and other members of his patriclan (fam) after staying up during an all-night house-raising in Longgar. After the actual raising, shortly before dawn, the women and men began to speak of what they subsumed under the rubric of *sejara* (or *sejarah*), a cognate of the Indonesian word for 'history,' about which I will have more to say later. It was in this setting that Pukulgore's name first came up, and as I was being given an overview of the particular 'history' (*sejara*) with which he is associated, several of this man's descendants wept.

According to Lakulu, the reason motivating Pukulgore's arrest by the Dutch also explains why the names of the leaders of the 'rebellion' had not been recorded by the colonisers on the back of a photograph that I had found in a Dutch archive entitled 'Heads of the Rebellion in the Aru Islands. Died in prison in Ambon, circa 1890' (see Figure 5).

Figure 5
Leaders of the revolt on the Aru Islands. Died in prison on Ambon, about 1890.[9]

9 Source: KITLV Leiden.

It was Pukulgore's powerful speech—at least partly attributable to this Longgar headman's possession of *ilmu*, or 'specialised knowledge'—that caused his arrest. In both Lakulu's and Belwi Kobawan's (a granddaughter of the leader) attribution to their kinsman of *ilmu*, this special knowledge is unburdened by the usual connotations of illegitimacy and covertness with which it is commonly identified on Barakai. Here, the presence of such 'black magic' is betrayed only by the bracelet adorning the leader's arm in the photograph—which Lakulu said was of solid gold—and by the powerful effects and transformations that were allegedly occasioned in the world whenever he spoke. Lakulu claimed that the effects of Pukulgore's powerful speech not only explained why the Dutch arrested him, but also why they immediately imposed silence on him once they did. In forbidding Pukulgore and the other Aruese leaders of the rebellion to speak, however, this meant that even their names were never voiced. This is why, following Lakulu, none of the leaders' names were ever recorded on the back of the archival photograph, as they were simply never spoken.

It is especially this moment of silence, or silencing, that the song commemorates at the very same time that it evokes a soundscape of considerable complexity. Following Lakulu's four-part division, the first part of the song tells how the moment of conception of the composition of the song corresponds to the moment at which the community's speech is broken and bereft of its force. What is more, the song is not even sung, but silently composed *a gul*, or 'in the head.' If you recall: 'he began to compose in his head the song [about the affair] that would break the village's powerful speech.' Quite striking here is that everything happens as if the song's composition, and therewith the emergence of *sejara*, actually anticipates the shattering of powerful speech. In other words, as modes of communication, *sejara* and powerful speech are somehow antithetical. In the song's second part, two Ambonese in the Dutch colonial service blow whistles as Pukulgore, the chief, begins to speak. In the third part, we are told that it is 'because of words spoken' that the Longgar chief is escorted by his grandson, Sirgwa, to the Dutch battleship. As the Aruese leader is led away, he hushes and imposes silence on his daughter, Magar, thereby generalising his own silencing by the colonial authorities. Yet, in the fourth and final part of the song—indeed, in its very last line—we are told that even as the Longgar chief is held and captured in the eyeglass of an Ambonese in the Dutch colonial service, Pukulgore is laughing a great laugh. Deprived of both his voice and name, and wholly suspended in the colonial gaze, the chief nevertheless laughs a great laugh.

This song differs in a number of important respects from other songs sung by Barakai women and men that are held to bear testimony to and be saturated with 'history,' or *sejara*. Unlike any other song that I know of, this one documents the moment of its own creation—which, more than anything else, is a moment of silence in which powerful words are broken and a song is composed without being sung. Yet, if the names and voices of Barakai islanders and their leaders are silenced, this moment of silence itself has a name: it is called *Zaman Belanda*, or the 'Dutch Era,' and it marks a break between the past of *sejara* and the present which, at the time I became acquainted with the song, was still that of the former New Order.

To understand the anomaly of the *Zaman Belanda* song, a brief and more general discussion of the interrelationships between songs and *sejara* on Barakai is necessary. Songs that can be brought to bear on *sejara* are classified locally as 'proofs,' due to their ability to serve as a material witness to a given 'history.' The song is held to be entangled with *sejara* while also commemorating it, such as the above song does to the *Zaman Belanda*.[10] This explains the off-hand comment on the night of the house-raising in Longgar when I was first told of Pukulgore's *sejara* that 'it has its song' (B.*ai ken sab*). In its capacity to witness, as well as codify, a particular repertoire of sentiments associated with a given *sejara*,[11] a 'proof'—whether in the form of a song, a special object such as a colonial staff office (D.*rottingsknop*), or a feature of Barakai's natural scenery—has an objectified, thing-like character. In the hands of a skilled storyteller who rallies the various 'proofs' that, familiar to her audience, texture and imbue her tale with the appropriate authority and sentiment, a given *sejara* acquires a certain objectivity. *Sejara* should therefore be understood not only as designating a 'history,' but also as the very referent about which a history is told—not only, then, as a discourse, but also as the discoursed world.

Besides the general silence both surrounding and imbuing the *Zaman Belanda* song, its most striking aspect is its idiosyncratic division into four parts. Songs on Barakai commonly consist of three as opposed to four parts—namely, a trunk, a tip, and an unmarked or lesser-marked middle. The ordering of songs from 'trunk' to 'tip' evokes their inherent futurity and the built-in capacity for songs to expand beyond themselves by generating distinct but related forms of narrative discourse. This particular understanding of songs is what allows them to be deployed in the projects of Barakai women and men—in discursive relation to other narrative forms—as an authorising source for their different actions, assertions, and claims.

The *Zaman Belanda* song with its unique four-part structure seems to foreclose the capacity inherent in other Barakai songs to 'grow,' as it were, and expand beyond themselves. Rather than futurity and growth, the four-part structure of the *Zaman Belanda* song suggests the kind of completeness that is associated with things fourfold on Barakai, something not granted to other songs on the island. The expression 'to take four,' or *dai kai*, for instance, often heard in ritual settings, means to encompass all four wind directions, and in doing so, to trace out a circle, thereby imaging a certain closure. In contrast to the futurity and open-

---

10 Spyer 2000; cf. Hoskins 1993: 120; Keane 1997.
11 See also Schulte Nordholt, this volume, for a discussion of the relation between different genres of representations of the past and sentiments.

endedness built into other songs, the only future the Zaman Belanda song anticipates is the shattering of words coincident with this song's own emergence. What is not at issue here is a *sejara*, which in its very structure intimates its possible applications in the projects of persons and collectivities. What arguably is at issue, rather, is a signpost or border marker constituting a break between a distinct closed-off 'Dutch Era' and a present social order, with respect to which the short-circuited *sejara* of a Zaman Belanda could have no relevance or bearing. As the song itself suggests, it documents not the telling or possibilities of history as *sejara*, but rather the radical suppression of powerful words.

This is presumably why the only point in the Zaman Belanda song when a potential singer comes to fully embody the *sejara* commemorated by the song (which, as you may have noticed, shifts perspective several times), or the only point when the action spoken of in the song occurs in the first person, is when the Longgar chief has already been silenced—his voice, name, and body arrested—and he boards the Dutch warship ('I climb up, I climb to the top of the ship,' and so forth). The 'I' of the song, which is the 'I' of the history of the rebellion, as well as the 'I' of Zaman Belanda is, therefore, a suppressed and silenced 'I,' or an arrested speaker in the full sense of that word. What complicates this reading somewhat is Pukulgore's great laugh caught in the colonial gaze as the last image of the song. It is plausible to see this conclusion as signalling the inability of the colonial state or, for that matter, its subsequent successors to impose a total hermetic silence on their subjects. At the same time, the meaning of this laugh is probably less self-evident than an interpretation of, for instance, 'native resistance' would imply. I prefer, therefore, to leave the status of the laugh, and therewith the question as to what kind of intention the song ascribes or does not ascribe to Pukulgore as an historical subject, open. Less spectacular than any confirmation of a clear-cut 'resistance,' what I believe the laugh conveys is how subjectivity can never be fully caught or stabilised, and necessarily remains in excess of the attempts to contain it. As a song which codifies the shattering of words and the interruption of discourse at a particular historical moment, it was, however, only appropriate that the song 'Zaman Belanda' was presented to me, in a manner that inscribed its own silence: already written, already codified as *sejara*, on a white piece of paper, and apparently linked to no particular project—except perhaps my own.

Having called attention to some of the idiosyncrasies of the Zaman Belanda song, I now propose an explanation as to why this song is so unlike any other that I encountered in Aru. Recall the extent to which both the concept of *sejara* and of a Zaman Belanda are inflected by the discourse of national historiography so that, even in such a frontier space of the nation-state as Aru, they may bear the traces of this wider Indonesian provenance. Let us first look, therefore, to this historiography to see if it offers anything which might help account for the anomaly of Aru's 'Dutch Era' and its identification with a violent silencing. In a recent, and remarkable, paper entitled 'Memory Work and Colonial Studies:

Recasting the Colonial in Contemporary Java,' Stoler and Strassler record their initial surprise at the reluctance and evasiveness of many Javanese to speak about the Dutch colonial past. As opposed to the 'safe,' highly codified public discourse on the Japanese occupation with its reiterated themes of suffering and sacrifice, accounts and memories of the 'three hundred years of Dutch rule' emerged with difficulty in a marginalised and fraught discursive terrain.[12] Beyond the officially codified 'three hundred' or 'three hundred and fifty' years of colonial oppression, there seemed to be no easily recoverable or even available memory sites. Even among those Javanese subalterns who, one might have thought, would have done the most to preserve the intimate recollections of their former colonisers—namely, those women and men who had worked as servants in Dutch colonial homes—were only fragmented rememberings. This 'thinness of public discourse about colonial oppression' stands in stark opposition to the anti-imperialism of the Sukarno period from which, by calling itself the 'New Order,' the Suharto regime aimed to distinguish itself since its inception.[13] According to Stoler and Strassler, such anti-imperialist sentiments evoked for the Suharto regime the 'extreme' leftism of the Sukarno period. As such, they were also regarded as likely to inhibit foreign investment and tourism.[14] What, invoking Pemberton, these authors quite rightly single out as predominantly accounting for the relative silencing of the colonial period under Suharto is 'the New Order state's own eerie resemblance to it.'[15]

The emphases, 'blind spots,' and repressions of nationalist historiography under Suharto reached Aru's Backshore in a complex manner through diverse channels such as the national language, Bahasa Indonesia, the country's educational system, the government and church officials who visited the archipelago's eastern pearldiving areas, traders who lived on the island, as well as through the national media—radio, and only very recently in Aru, national television. These kinds of influences are, of course, difficult to track. What seems clear, however, is that the relative marginalisation of *Zaman Belanda* in the New Order's official historiography has indeed left an indelible trace in Barakai's own Dutch Era song.

Be this as it may, under the political conditions of the New Order, the song's theme already justified its own careful hedging by silence, or by the short-circuiting of the futurity otherwise commonly built into the very structure of Barakai songs. This theme also explains why the 'I' of the song may only emerge under arrest, as the signifier of a confined and silenced subject. This theme is of a displaced insurrection as well as of the political effects which may be brought about should potent words be released and set into circulation. Everything that

---

12 Stoler and Strassler 1998: 11-12.
13 Ibid.
14 Ibid.
15 Stoler and Strassler 1998; cf. Pemberton 1994.

has happened in Indonesia since the stepping down of Suharto in May of 1998 suggests, indeed, how the circulation of rumours or powerful words such as 'reformasi' can set off momentous happenings and have profound social and political consequences. However, at the time I was handed the Zaman Belanda song on a white piece of paper in the late 1980s, the possibility of the kind of political action and aspirations that we have witnessed in the last year or an end to the New Order (if that is what it is) seemed extremely remote. For this reason, at that time, the power that Barakai islanders commonly attributed to sejara could not—when it came to the Zaman Belanda song—be accessed in the usual manner. In contrast to other songs sung as 'proofs' to authorise the deployment of sejara in different situations, the colonial history commemorated in the Zaman Belanda song could, following Lakulu, only be tapped at the grave of his ancestor and, crucially, only by engaging in ilmu. It is this exclusive mediation of Pukulgore's formidable powers by ilmu which betrays how calling up the sejara of the insurrection against colonial authority was felt to be charged as a politically suspect act tainted with illegitimacy under New Order conditions.

The relative silence in New Order historiography on the Dutch Era—foreclosing any easy comparison between that past and the oppressive Suharto present—together with the mobilisation of persons against the powers-that-be by 'words spoken,' suggests why a radical split necessarily severed the words of the past from the violently imposed silence of the recent Suharto present. A period unto itself, Zaman Belanda stands as a signpost between powerful Aru leaders like Pukulgore with their ability to sway entire populations and the political illegitimacy of alternative visions, voices, and authorities under Suharto's hard-handed rule.

Pocketbook Religiosity

In one distant pre-religious past, according to a story sometimes told on Barakai, the difference between Muslims and pagans emerged out of an accident of dress, following, that is, from the apparently trivial circumstances of having or not having a pocket. Familiar in its characteristic synoptic form across the island, the story zooms in on the decisive moment when two Aruese ancestors—one wearing a so-called openjas (or 'jacket,' as derived from the Dutch), and the other a loincloth—were offered a copy of the Qu'ran. As if religious difference extends directly from distinctions of dress, the openjas, being graced with a pocket, immediately accommodated a fit between this Aruese ancestor and the new faith.[16] By the same token, the other Aruese ancestor, with no place to put even a pocket-size Qu'ran, remained pagan by default of his dress.

16 Cf. Pemberton 1994: 132-136.

The beauty of this story lies in its simplicity and radical telescoping into a single scene of a tumultuous time of coercion and violence. I can only sketch here the outlines of this history and suggest how many of its most salient dimensions for Aruese are condensed in the stark outlines of the above tale. The absolute turnabout in the story—hinging entirely on having or not having a pocket—institutes a radical break between those who embrace religion and those who are excluded by it. Significantly, this absolute split reiterates the Suharto regime's own religious politics and, in particular, the marked divide this regime posited between those persons who, as it was put, did 'not yet have a religion' (I.*belum beragama*), versus those who 'already' did (I.*sudah beragama*). The tale further hints at a temporality which is a frequent legacy of missionisation: the split in mission historiography between a before- and after-conversion. Most importantly, however, the tale of the pocket, in one single powerful image, collapses the abrupt violence of a transformation in identity that was imposed from without in the form of a simple, if dramatic, shift in 'outward appearance.'[17] It is indeed telling that in the tale, this shift takes on such a material form and, more specifically, is encoded as a dramatic difference in types of clothing.

Importantly, the above synopsis of *agama*'s introduction is always referred to as the arrival at Aru of religion *in general*, rather than of Islam specifically. Indeed, this is already implied insofar as the pivotal contrast on which the story turns is that of pagans and converts. It thereby exposes the dynamics of a religious politics which drew hard and fast lines between those Indonesians with *agama* and those without. This religious politics also separated Muslim, Protestant, Catholic, Hindu, and Buddhist national citizens, or those adhering to one of the five religions recognised by the Suharto government, out from a residual, recalcitrant, and undifferentiated pagan body. Since the mass conversion of Aruese pagans orchestrated by the archipelago's civil and military officialdom in 1976-77—motivated by the determination to ensure the participation of all islanders in the 1977 national elections, with participation, in turn, contingent upon possessing a citizen's identity card on which the category of *agama* had to be declared—all Backshore peoples 'already have a religion.'[18] The implications of this double conversion to *agama* and citizenship, of crossing the gap between the 'not yet' and the 'already,' are far-reaching. As Kipp and Rodgers observe, the close connection posited under the New Order among *agama*, good citizenship, and progress meant that, by following 'an implicit logic of opposites,' those persons and collectivities who remained in a state of default—without *agama*, or, as in the story, a pocket—were, by implication, backwards, uncommitted to the values of the state ideology (*Pancasila*), and disloyal national citizens.[19]

---

17 Schulte Nordholt (ed.) 1997.
18 Spyer 1996.
19 Kipp and Rodgers 1987: 23.

I have discussed the explicit materiality in which violence is dressed in the tale elsewhere.[20] Suffice it to say here that in Aru, as in many other places, religion, clothing, and coercion have long been closely linked. Since the 1920s on, when the first Protestant school teachers came from Ambon to the Backshore, they intervened materially as much as spiritually in the refashioning of pagan Aruese. As elsewhere, some of the most violent assaults on local appearance aimed at restructuring male and female attire along proper Christian and gendered lines.[21] Beyond the usual covering over of especially female bodies, some older men on Barakai still recalled having their earrings ripped right out of their lobes by their Protestant schoolteacher. In the early 1960s, when Dutch Catholic missionaries first began their own proselytisation on the Backshore, they too set about re-figuring some of the material grounds on which the conversions of the mid-1970s would subsequently be played out—as they brought not only medicine and schooling, but notably more proper Christian attire to Barakai communities.

As the drive for the religious conversion of Aru's pagans assumed momentum in the mid-1970s, local representatives of the Protestant Church of the Moluccas, the Dutch Sacred Heart Mission, and Islam competed vigorously for the religious affiliation of pagan Aruese. Aruese memories of this time foreground the commodities with which persons were seduced and bribed into one or another faith; they also emphasise the complete lack of alternatives available to them within the coercive and often violently executed decision of the district office in Dobo to impose on every last pagan Aruese an officially recognised *agama*. In the clear-cut abruptness of the transformation it describes, the story of the pocket depicts in stark relief the lack of a fit between a pagan body and foreign *agama*cised clothes. Therewith, it hints at the violence which left Aruese pagans vulnerable and exposed to the glaring gaze of the State in the context of their coerced conversion—much as in the biblical story in which Adam and Eve simultaneously discover their nudity and their shame. It also hints at the close identification in this conversion setting between clothing and—in the case of Bemun where I resided—Catholicism. Thus, the story posits religion as the only material place which offers, pocket-like, a place to hide. As one instance of a splitting historiography, the absolute rift between pagans and converts and the radical turnabout in the pocket's tale register, in stark Manichean set-up, the dramatic application of the state's religious politics on the Backshore, the materiality of religion's introduction to Aru, and the violence with which the double coercion to *agama* and citizenship within the nation-state was imposed on Barakai peoples in the mid-1970s.

---

20 Spyer 1999; 2000.
21 Comaroff and Comaroff 1991-1997.

Coup de Rituel

In this last example, I look briefly at the break between the past of the ancestors and the present of contemporary Barakai, that is, the time at which the women and men of the pearldiving community of Bemun open their annual ritual performance. My objective here is to point out that the break constituting this performance's beginning takes the form of a surprise assault or 'coup' in which the past appears to overtake and displace the present, thereby introducing the possibility of an ancestral space-time into the Barakai everyday. The performance's beginning hinges on the effect of a specific theatricality, which necessarily involves an element of surprise. Rather than a simple 'raising of curtains' inaugurating the performance, at issue is the active production of the performance's very start, and therein, the inauguration of an 'other' space-time.

Held at the seasonal transition from the east to the west, Bemun's performance opens the season for collecting pearl oysters and other sea products for trade and, relatedly, for intensive commerce with the ethnic Chinese traders on the island. Given the seasonal nature of this trade, climatic criteria such as the condition of the Barakai seas have always been critical to the opening of the performance. The timing of Bemun's performance should also be co-ordinated in relation to those held by the three neighbouring communities on the island. Other temporal criteria are also relevant and need to be taken into account by the paired officials, the Prow and the Stern, who together organise the main events of the performance. Since it takes place sometime between mid-September and early October, these officials must, for instance, take care to avoid any overlap with the celebration of the National Armed Forces day on October 5, since 'government' together with religion, trade, and many other things are rigorously excluded from the space of the performance. In various practical ways, such as setting aside stores of food for the feast, Bemunese villagers must also have readied themselves, their homes, and their gardens for the demand of their annual celebration. Given that the first rite of the performance is a fish-poisoning held at a sacred estuary in which the participants stand for several hours, the opening of the performance as a whole must take place when low tide occurs in the relatively cool morning.

With varying emphases for any given performance, a plethora of conditions (such as calendars and other considerations which are too numerous to mention here) figure as a loose frame within which the Prow—whose speciality is initiating ritual action—ideally makes the final decision regarding when to begin Bemun's performance. Year after year, as far as I could tell from talking to Bemunese, this collective framework scales down the possibilities almost, but not quite, to the day. Within these parameters, an element of surprise is expected and even relished. Indeed, the surprise provoked by the Prow's decision to 'walk' one evening and set it all in motion—following, in both 1986 and 1987, a surplus of talk and speculation regarding the performance's precise beginning—clearly contributed much to this same beginning's efficacy. Since the Prow's 'walk' or round of the village should occur unseen and uninterrupted by encounters with his fellow villagers—'invisibly,' as it were—and since he is escorted by ancestral spirits, the surprise of this opening has a slightly unsettling aspect to it, a feeling that significantly contributes to its power.

Because the beginning is 'invisible' to start with, the opening act of the ritual season is never entirely predictable and seemed to take most Bemunese somewhat aback, not the least even the Stern with whose house the Prow opens his round. In expectation that the Prow might walk, villagers keep to their homes after dark around the time when the performance is likely to begin. It is only when the Prow appears framed in the doorway of their homes as a dark silhouette in ancestral dress that they are confronted with their performance's beginning and with the space-time it introduces. Ideally, with the suddenness with which the night overcomes the day in the equatorial tropics, the Prow emerges in the loincloth of his ancestors and makes the round in which he summons Bemun's men to gather at the village centre. In this way, his entry has the effect of a minor 'coup' that transgresses and disrupts the more mundane space of circulation in which he and his fellow Bemunese move, thus making way for something radically different.[22] In Michel de Certeau's terms, this kind of artful ploy '*makes* a hit ('coup') far more than it simply describes one; its force lies in the effect it has of seeming to elude present circumstances through the creation of a fictional space—a space, as it were, of 'once upon a time there was...'[23] The fact that the beginning of the performance occurs as a disruption in which the past overtakes the present also accounts for its perceived danger and the association of this danger with the ancestral spirits who accompany the Prow on his round. While it is only the beginning of a performance which brings about a radical break between an 'Aru' past of the ancestors and a 'Malay' present of trade, missionisation, and the nation-state, its uncanny opening contains an implicit violence and, as with the other split historiographies considered here, is therefore especially charged.

In these three examples, I have focused on the ways in which national historiography and local history-making intersect with complex results. Thus, if Aru's splitting historiographies locally re-edit the break through which the New Order established its authority by setting itself up as the telos of a pagan past left behind, this is not all that they do. As I have suggested, especially in the last example, this splitting may, at times, also invest the past with a local force which, although ambivalent, can serve as an authorising source for the present.

---

22 De Certeau 1984: 79.
23 De Certeau 1984.

In other words, Aru's splitting historiographies are an important means through which Aruese negotiate their place in one of the frontier spaces of the nation-state. I use the term 're-editing' advisedly to characterise Aru's splitting historiographies. On the one hand, the three examples offered above show how Aruese rework and re-deploy those tokens of national historiography which tend to come to the island as broken-off-bits of a more monumentalised state project. Yet, by using such fragments to assemble history, and remaining, thereby, somehow indebted to the 'official' structures of the regime, Backshore women and men also shape an Aruese 'historiography' of sorts—here understood loosely as simply the making and codification of historical knowledge. If, in some ways, this history-making consolidates their subordinate position within a national hierarchy, it becomes, in other ways, a way of seizing their own capacity for change and agency within it.

References

Adam, Barbara, *Time and Social Theory* (Cambridge: Polity Press, 1990).
Adjaye, Joseph K., 'Time in Africa and Its Diaspora: An Introduction,' in Joseph K. Adjaye (ed.), 1994, 1-16.
Adjaye, Joseph K. (ed.), *Time in the Black Experience* (Westport, Conn., and London: Greenwood Press, 1994).
Amin, Shahid, and Dipesh Chakrabarty (eds.), *Subaltern Studies IX: Writings on South Asian History and Society* (Delhi: Oxford University Press, 1996).
Asao Naohiro, Kano Masanao et al. (eds.) *Nihon tsûshi, bekkan 1: Rekishi ishiki no genzai* (Tokyo: n.p., 1995).
Augustine, *Confessions*, Henry Chadwick (trans.) (Oxford: Oxford University Press, 1991).
Aveni, Anthony, *Empires of Time: Calendars, Clocks, and Cultures* (London: I.B. Tauris, 1990).
Baczko, Bronislaw, 'Le calendrier républicain: décréter l'éternité,' in: Pierre Nora (ed.), 1997, Vol. I, 67-106.
Baetghe, Martin, and Wolfgang Essbach (eds.), *Soziologie, Entdeckungen im Alltäglichen: Hans Paul Bahrt, Festschrift zu seinem 65 Geburtstag* (Frankfurt: Campus, 1983).
Barrow, John D., *The Theories of Everything: The Quest for Ultimate Explanation* (Oxford: Clarendon Press, 1991).
Bateson, G., 'An Old Temple and a New Myth,' in J. Belo (ed.) 1970, 111-136 [1937].
Belo, J. (ed.), *Traditional Balinese Culture* (New York: Columbia University Press, 1970).
Bernot, Lucien, *Les paysans arakanais du Pakistan oriental: l'histoire, le monde végétal et l'organisation sociale des réfugiés Marma (Mog)* (Paris and The Hague: Mouton, 1967).
Bhattacharjee, J.B. (ed.), *Proceedings of the North East India History Association, Sixth Session, Agartala, 1985* (Shillong: North Eastern Hill University, 1986).
Bitô Masahide, 'Kôkoku shikan no seiritsu,' in Sakara, Bitô and Akiyama (eds.), 1984, 299-348.
Blacker, Carmen, *The Japanese Enlightenment* (Cambridge: n.p., 1954).
Bloch, Marc, *Feudal Society*, 2 vols. (London: Routledge and Kegan Paul, 1961-62).
Bloch, Maurice, 'The Past and the Present in the Present,' *Man*, 12 (1977), 278-292.

Boorstin, Daniel J., *De ontdekkers: de zoektocht van de mens naar zichzelf en zijn wereld* (trans. from English, 1983) (Amsterdam: Elsevier, 1991).

Burke, Peter, *The French Historical Revolution. The Annales School, 1929-89* (Cambridge: n.p., 1990).

Campbell, Jeremy, *Winston Churchill's Afternoon Nap: A Wide-Awake Inquiry into the Human Nature of Time* (New York: Simon and Schuster, 1986).

Cipolla, Carlo M., *Clocks and Culture 1300-1700* (London: Collins, 1967).

Cohen, John, 'Subjective Time,' in J.T. Fraser (ed.), 1968, 257-275.

Cohn, Bernhard S., 'The Pasts of an Indian Village,' in Diane Owen Hughes and Thomas R. Trautmann (eds.), 1995, 21-30 (originally published in *Comparative Studies in Society and History*, 3:3 [1961], 241-249).

Comaroff, John L., and Jean Comaroff, *Of Revelation and Revolution: Christianity, Colonialism, and Consciousness in South Africa* (Chicago: University of Chicago Press, 1991-1997), 2 Vols.

Creese, H., 'Balinese Babad as Historical Sources: A Reinterpretation of the Fall of Gelgel,' *Bijdragen tot de Taal-, Land- en Volkenkunde*, 147 (1991), 236-260.

Creese, H., 'Chronologies and Chronograms: An Interim Response to Hagerdal,' *Bijdragen tot de Taal-, Land- en Volkenkunde*, 151 (1995), 125-131.

Creese, H., 'Palm Leaves and Paraboles: Contemporary Historical Discourse in Bali,' paper for the IIAS Workshop The Pace of Life in Southeast Asi, Leiden, November 19-22, 1997.

Cribb, Robert, 'Jam Karet: Telling the Time in Indonesia,' paper for the NIAS Conference Time and Society in Modern Asia, Copenhagen, June 18-20, 1998.

De Certeau, Michel, *The Practice of Everyday Life* (Berkeley: University of California Press, 1984).

Dodd, W.C., *The Thai Race: Elder Brother of the Chinese* (Bangkok: White Lotus Press, 1996).

Dohrn-van Rossum, Gerhard, *History of the Hour: Clocks and Modern Temporal Orders* (Chicago: University of Chicago Press, 1996).

Duara, Prasenjit, *Rescuing History from the Nation: Questioning Narratives of Modern China* (Chicago and London: n.p., 1995).

Duara, Prasenjit, 'The Regime of Authenticity: Timelessness, Gender, and National History in Modern China,' *History and Theory*, 37 (1998), 287-308.

Eaton, Richard M., *The Rise of Islam and the Bengal Frontier, 1204-1760* (Berkeley: University of California Press, 1993).

Elias, Norbert, *Time: An Essay* (Oxford: Basil Blackwell, 1992).

Elias, Norbert, *The Civilizing Process* (Oxford: Basil Blackwell, 1994).

Ember, A.T., and C. Gluck (eds), *Asia in Western and World History* (Armonk/London: M.E. Sharpe, 1997).

Escobar, Arturo, *Encountering Development: The Making and Unmaking of the Third World* (Princeton, N.J.: Princeton University Press, 1995).

Evans-Pritchard, E.E., *Witchcraft, Oracles and Magic Among the Azande* (Oxford: Clarendon, 1937).

Evans-Pritchard, E.E., *The Nuer* (Oxford: Oxford University Press, 1940).

Fabian, Johannes, *Time and the Other: How Anthropology Makes Its Object* (New York: Columbia University Press, 1983).
Featherstone, Mike, Scott Lash, and Roland Robertson (eds.), *Global Modernities* (London, etc: Sage, 1995).
Ferrand, Gabriel (ed. and transl.), *Rélations de voyages et textes géographiques arabes, persans et turks rélatifs à l'Extrême-Orient du VIIIe au XVIIIe Siècles*, vol. 2 (Paris: Ernest Leroux, 1914).
Fogel, Joshua A., 'The Debates over the Asiatic Mode of Production in Soviet Russia, China, and Japan,' in Joshua A. Fogel (ed.), 40-65.
Fogel, Joshua A. (ed.), *The Cultural Dimension of Sino-Japanese Relations. Essays on the 19th and 20th Centuries* (New York: n.d., 1995).
Flood, E. Thadeus, 'Sukhothai-Mongol Relations,' *Journal of the Siam Society*, 57:2 (1969), 201-258.
Fraser, J.T. (ed.), *The Voices of Time: A Cooperative Survey of Man's Views of Time as Expressed by the Sciences and by the Humanities* (London: The Penguin Press, 1968).
Geertz, Clifford, *Person, Time and Conduct in Bali: An Essay in Cultural Analysis* [Cultural Report Series 14] (Hartford, Conn.: Yale University, Southeast Asia Studies, 1966).
Geertz, Hildred (ed.), *State and Society in Bali: Historical, Textual and Anthropological Approaches* (Leiden: KITLV Press, 1991).
Gleichmann, Peter Reinhart, 'Nacht und Zivilisation,' in Martin Baetghe and Wolfgang Essbach (eds.), 1983.
Goi Naohiro, *Kindai Nihon to toyôshigaku* (Tokyo: n.p., 1976).
Goris, R., 'Holidays and Holy Days,' in J. Swellengrebel (ed.), 1960, 115-129.
Gosden, Christopher, *Social Being and Time* (Oxford: Blackwell, 1994).
Goudsblom, Johan, *Sociology in the Balance* (Oxford: Basil Blackwell, 1977).
Goudsblom, Johan, *Het regime van de tijd* (Amsterdam: Meulenhoff, 1997).
Goudsblom, Johan, E.L. Jones, and Stephen Mennel, *The Course of Human History* (Armonk, NY: M.E. Sharpe, 1996).
Grebing, Helga, et al., *Der 'deutsche Sonderweg' in Europa. 1806-1945. Eine Kritik* (Stuttgart: n.p., 1986).
Greenough, Paul R., *Prosperity and Misery in Modern Bengal: The Famine of 1943-1944* (New York, etc.: Oxford University Press, 1982).
Grew, Raymond, 'Foreword,' in Diane Owen Hughes and Thomas R. Trautmann (eds.), 1995, vii-xi.
Grimm, T., 'Thailand in the Light of Official Chinese Historiography: A Chapter in the 'History of the Ming Dynasty',' *Journal of the Siam Society*, 49:1 (July 1961), 1-20.
Griswold, A.B., and Prasert na Nagara, 'A Law Promulgated by the King of Ayudhya in 1397 AD,' *Journal of the Siam Society*, 57:1 (Jan. 1969), 109-148.
Griswold, A.B., and Prasert na Nagara, 'King Lödaiya of Sukhodaya and His Contemporaries, Epigraphic and Historical Studies, No. 10,' *Journal of the Siam Society*, 60:1 (January 1972), 21-152.
Griswold, A.B., and Prasert na Nagara, 'Kingship and Society at Sukhodaya,' in G.W. Skinner and A. Thomas Kirsch (eds.), 1975, 29-92.

Guha, Ranajit, 'The Small Voice of History,' in Shahid Amin and Dipesh Chakrabarty (eds.), 1996, 1-12.

Hägerdal, H., 'Bali in the Sixteenth and Seventeenth Centuries: Suggestions for a Chronology of the Gelgel Period,' *Bijdragen Tot de Taal-, Land- en Volkenkunde*, 151 (1995a), 101-124.

Hägerdal, H., 'Reply to Dr. Helen Creese,' *Bijdragen Tot de Taal-, Land- en Volkenkunde*, 151 (1995b), 292-293.

Harber, Hans, and Sjaak Koenis (eds.), *Kennis en Methode: Special Issue on Sociale Cohesie, Cultuur en de Dingen* (n.p., 1999).

Herzfeld, Hans, 'Menschenrecht und Staatsgrenze in Zwischeneuropa,' *Ausgewählte Aufsätze* (Berlin: n.p., 1962), 216-228.

Hobart, Mark, 'The Missing Subject: Balinese Time and the Elimination of History,' *RIMA*, 31 (1997), 123-172.

Hooker, Virginia Matheson (ed.), *Culture and Society in New Order Indonesia* (Oxford: Oxford University Press, 1993).

Horigome Yôzô, *Rekishi to ningen* (Tokyo: n.p., 1965).

Hoskins, Janet, *The Play of Time: Kodi Perspectives on Calendars, History, and Exchange* (Berkeley: University of California Press, 1993).

Hoston, Germaine A., *Marxism and the Crisis of Development in Prewar Japan* (Princeton: n.p., 1986).

Howe, Leo, 'The Social Determination of Knowledge: Maurice Bloch and Balinese Time,' *Man*, 16 (1981), 220-234.

Howse, Derek, *Greenwich Time and the Discovery of the Longitude* (Oxford: Oxford University Press, 1980).

Hughes, Diane Owen, and Thomas R. Trautmann (eds.), *Time: Histories and Ethnologies* (Ann Arbor: University of Michigan Press, 1995).

Huizinga, Johan, *Hoe bepaalt de geschiedenis het heden? Een niet-gehouden rede* (Haarlem: Tuindorp/Boom-Ruygrok, 1946).

Ienaga Saburô, *History of Japan* (Tôkyô: n.p., 1953).

Ienaga Saburô, 'Keimô shigaku,' in Matsushima Eiichi (ed.), 1954, 422-427.

Inoue Kiyoshi, *Nihon no rekishi*, vol.1 (Tokyo: n.p., 1963).

Jahangir, Burhanuddin Khan, *The Quest of Zainul Abedin* (Dhaka: International Centre for Bengal Studies, 1993).

Jahangir, Burhanuddin Khan, 'Sonar Bangla: A Political Myth Reconstruction of History and Politics in Bangladesh,' *Journal of Social Studies*, 80 (April 1998), 1-9.

Karatani Kôjin, 'The Discursive Space of Modern Japan,' in Masao Miyoshi and Harry D. Harootunian (ed.), 1993, 288-315.

Kan San Jun, 'Ajiakan no sôkoku. Shokuminchi shugisha no sozô,' *Seikyu*, 22 (1995), 166-171 and 24 (1995), 101-105.

Kan San Jun, *Orientarizumu no achira e. Kindai bunka hihan* (Tokyo: n.p., 1996).

Keane, Webb, *Signs of Recognition: Powers and Hazards of Representation in an Indonesian Society* (Berkeley: University of California Press, 1997).

Kern, Stephen, *The Culture of Time and Space 1880-1918* (Cambridge, Mass.: Harvard University Press, 1983).
'Khon Thai mai dai ma chak nai?' (Where did the Thais come from?), in Suchit (ed.), 1984.
Kipp, Rita Smith, and Susan Rodgers, 'Indonesian Religions in Society,' in Rita Smith Kipp and Susan Rodgers (eds.), 1987, 1-31.
Kipp, Rita Smith, and Susan Rodgers (eds.), *Indonesian Religions in Transition* (Tucson: University of Arizona Press, 1987).
Knippenberg, Hans, and Ben De Pater, *De Eenwording van Nederland* (Nijmegen: Sun, 1988).
Kocka, Jürgen, 'German History before Hitler: The Debate about the German 'Sonderweg',' *Journal of Comparative History*, 23 (1988), 3-16.
Koloniaal Verslag (Den Haag: Algemeene Landsdrukkerij, 1893).
Koselleck, Reinhart, 'Geschichte, Geschichten und formale Zeitstrukturen,' *Vergangene Zukunft. Zur Semantik geschichtlicher Zeiten* (Frankfurt: n.p., 1979), 130-143.
Kumar, A., *Java and Modern Europe* (Richmond: Curzon, 1997).
Kuwabara Takeo, 'Tradition versus Modernization,' *Japan and Western Civilization. Essays on Comparative Culture* (Tôkyô: n.p., 1983), 39-64.
Laclau, Ernesto, and Chantal Mouffe, *Hegemony & Socialist Strategy: Towards a Radical Democratic Politics* (London and New York: n.p., 1985).
Leduc, Jean, *Les historiens et le temps: Conceptions, problématiques, écritures* (Paris: Éditions du Seuil, 1999).
LeGoff, Jacques, *Pour un autre Moyen Age: Temps, Travail, et Culture en Occident* (Paris: Gallimard, 1977).
Levine, Robert, *A Geography of Time: The Temporal Misadventures of a Social Psychologist, or How Every Culture Keeps Time Just a Little Bit Differently* (New York: Basic Books, 1997).
Lieberman, Victor B., *Burmese Administrative Cycles: Anarchy and Conquest, c.1580-1760* (Princeton, 1984).
Lindsey, Timothy C., 'Concrete Ideology: Taste, Tradition, and the Javanese Past in New Order Public Space,' in Virginia Matheson Hooker (ed.), 1993, 166-182.
Loknath Haph-Punjika 1405 Saner (Dhaka: Loknath Book Agency, 1998).
Lovric, B., 'Bali: Myth, Magic and Morbidity,' in N. Owen (ed.), 1987, 117-141.
Luce, G.H., 'The Early Siam in Burma's History,' *Journal of the Siam Society*, 46:2 (1958a), 123-214.
Luce, G.H., 'The Early Siam in Burma's History, a Supplement,' *Journal of the Siam Society*, 47:1 (1958b), 59-101.
Lyngdoh, M.P. Rina, 'The Khasi System of Calculating Time,' in J.B. Bhattacharjee (ed.), 1986, 115-125.
Ma Huan, *Ying-yai sheng-lan chiao-chu, the Overall Survey of the Ocean's Shores* [1433], transl. Feng Ch'eng-Chuen, ed. J.V.G. Mills (Cambridge: University Press for the Hakluyt Society, 1970).
Macey, Samuel L. (ed.), *Encyclopaedia of Time* (New York and London: Garland Publishing, 1994).

Marx, Karl, *Het Kapitaal: Een Kritische Beschouwing Over de Economie*, Deel 1. (trans. from German, 1867), 10th ed. (Bussum: De Haan, 1978).
Masao Miyoshi and Harry D. Harootunian (eds.), *Japan in the World* (Durham and London: n.p., 1993).
Matsushima Eiichi (ed.), *Meiji shiron shû 1: Meiji bungaku zenshû*, vol. 77 (Tôkyô: n.p., 1954).
Minshu shugi kagakusha kyôkai, Rekishigaku kenkyûkai (eds.), *Nihon no rekishi* (Tokyo: n.p., 1949).
Munn, Nancy D., 'The Cultural Anthropology of Time: A Critical Essay,' *Annual Review of Anthropology*, 21 (1992), 93-123.
Nagahara Keiji, 'Jidai kubunron,' in Rekishigaku Kenkyûkai and Nihonshi Kenkyûkai (eds.), 1971, 3-40.
Nagahara Keiji, *Kôkoku shikan* (Tokyo: n.p., 1983).
Nakamura, Hajime, 'Time in Indian and Japanese Thought,' in J.T. Fraser (ed.), 1968, 77-91.
Naruse Osamu, *Sekaishi no ishiki to riron* (Tokyo: n.p., 1977).
Nasreen, *100 Poems of Taslima ——*, Kabir Chowdhury (trans.) (Dhaka: Ananya, Agrahayan 1404 / December 1997).
Needham, Joseph, 'Time and Knowledge in China and the West,' in J.T. Fraser (ed.), 1968, 92-135.
Nora, Pierre, 'La génération,' in Pierre Nora (ed.), 1997, Vol. 2.
Nora, Pierre (ed.), *Les lieux de mémoire* (Paris: Éditions Gallimard, 1997), 3 vols.
Nowotny, Helga, *Time: The Modern and Postmodern Experience* (Cambridge: Polity Press, 1994).
Osmany, Shireen Hasan, *Bangladeshi Nationalism: History of Dialectics and Dimensions* (Dhaka: University Press Limited, 1992).
Osterhammel, Jürgen, *Die Entzauberung Asiens. Europa und die Asiatischen Reiche im 18. Jahrhundert* (München; n.p., 1998).
Östör, Ákos, *The Play of the Gods: Locality, Ideology, Structure and Time in the Festivals of a Bengali Town* (Chicago and London: The University of Chicago Press, 1980).
Östör, Ákos, *Vessels of Time: An Essay on Temporal Change and Social Transformation* (Delhi: Oxford University Press, 1993).
Owen, N. (ed.), *Death and Disease in Southeast Asia: Explorations in Social, Medical and Demographic History* (Singapore: Oxford University Press, 1987).
Pelliot, Paul, *Mémoires sur les coutumes du Cambodge. Par Tcheou Ta-kouan*, Traduits et annotés par Paul Pelliot. Extrait du Bulletin de l'Ecole Francaise d'Extrême-Orient, Vol. 2 (1902).
Pemberton, John, *On the Subject of Java* (Ithaca, NY: Cornell University Press, 1994).
'Phrarachaphongsawadan Krungsi'ayutthaya chabap Luang Prasoet,' *Prachumphongsawadan*, Vol. 1 (Bangkok: Khurusapha, BE 2506).
Picard, M., *Bali: Cultural Tourism and Touristic Culture* (Singapore, Archipelago Press, 1996).
Pieters, P., *Praktische Indische Tolk: Spreken in Hollandsch, Maleisch en Javaansch* (Amsterdam: Van Holkema en Warendorf, n.d.).

Pigeaud, Th., *Java in the 14th Century, A Study in Cultural History, The Nagara-Kertagama by Rakawi Prapa-ca of Majapahit, 1365 AD* (The Hague: Martinus Nijhoff, 1960).

Prigogine, Ilya, and Isabelle Stengers, *Order Out of Chaos: Man's New Dialogue with Nature* (London: Heinemann, 1984).

Rekishigaku Kenkyûkai (ed.), *Rekishi ni okeru minzoku no mondai: Rekishigaku kenkyûkai 1951 nendo taikai hôkoku* (Tokyo: n.p., 1951).

Rekishigaku Kenkyûkai, and Nihonshi Kenkyûkai (eds.), *Kôza Nihonshi 9: Nihon shigaku ronsô* (Tokyo: n.p., 1971).

Robinson, G., *The Dark Side of Paradise: Political Violence in Bali* (Ithaca and London: Cornell Univeristy Press, 1995).

Rockhill, W.W., 'Notes on the Relations and Trade of China with the Eastern Archipelago and the Coast of the Indian Ocean during the Fourteenth Century,' *T'oung Pao*, 16 (1915), 61-159.

Rotenberg, Robert, *Time and Order in Metropolitan Vienna* (Washington, DC: Smithsonian Institution Press, 1992).

Rowe, William T., 'Approaches to Modern Chinese Social History,' in Olivier Zunz (ed.), 1985, 236-296.

Said, Edward W., *Orientalism. Western Conceptions of the Orient* (London: n.p., 1991 [1978]).

Saitô Takashi, *Shôwa shigakushi nôto: Rekishigaku no hassô* (Tokyo: n.p., 1984).

Sakara Tôru, Bitô Masahide, and Akiyama Ken (eds.), *Kôza Nihon shisô*, Vol. 4: *Jikan* (Tokyo: n.p., 1984).

Sarkar, Jadunath (ed.), *The History of Bengal*, Vol. II: *Muslim Period, 1200-1757* (Dacca: The University of Dacca, 1948).

Schmied, Gerhard, *Soziale Zeit: Umfang, 'Geschwindigkeit' und Evolution* (Berlin: Duncker and Humblot, 1985).

Schonberger, Howard B., *Aftermath of War: Americans and the Remaking of Japan 1945-1952* (Kent: n.p., 1989).

Schreiner, K., 'The Making of National Heroes: Guided Democracy to New Order,' in Henk Schulte Nordholt (ed.), 1997, 259-290.

Schulte Nordholt, Henk, *State, Village and Ritual in Bali: A Historical Perspective* (Amsterdam: VU University Press [Comparative Asian Studies 7], 1991a).

Schulte Nordholt, Henk, 'Temple and Authority in South Bali 1900-1980,' in H. Geertz (ed.), 1991b, 137-163.

Schulte Nordholt, Henk, 'Origin, Descent and Destruction: Text and Context in Balinese Representations of the Past,' *Indonesia*, 54 (1992), 27-58.

Schulte Nordholt, Henk, 'The Making of Traditional Bali: Colonial Ethnography and Bureaucratic Reproduction,' *History and Anthropology*, 8 (1994), 89-127.

Schulte Nordholt, Henk (ed.), *Outward Appearances: Dressing State and Society in Indonesia* (Leiden: KITLV Press, 1997).

Shaffer, Linda, 'A concrete panoply of intercultural exchange: Asia in world history,' in A.T. Ember, and C. Gluck (eds), 1997, 810-866.

Shigakkai (ed.), *Shigaku bunken mokuroku 1946-1950* (Tokyo: n.p., 1951a).

Shigakkai (ed.), *Nihonshi gaikan* (Tokyo: n.p., 1951b).

Shiozawa Kimio, *Ajiateki seisan yôshiki ron* (Tokyo: n.p., 1970).
Skinner, G.W., and A. Thomas Kirsch (eds.), *Change and Persistence in Thai Society* (Ithaca: Cornell University Press, 1975).
Skocpol, Theda, *States and Social Revolutions: A Comparative Analysis of France, Russia and China* (Cambridge: Cambridge University Press, 1979).
Sobhan, Rehman, *The Crisis of External Dependence: The Political Economy of Foreign Aid to Bangladesh* (Dhaka: University Press Ltd., 1982).
Spyer, Patricia, 'Serial Conversion/Conversion to Seriality: Religion, State, and Number in Aru, Eastern Indonesia,' in Peter van der Veer (ed.), 1996, 171-198.
Spyer, Patricia, 'What's in a Pocket?: Religion and the Formation of a Pagan Elsewhere in Aru, Eastern Indonesia,' in Hans Harber and Sjaak Koenis (eds.) 1999.
Spyer, Patricia, *The Memory of Trade: Modernity's Entanglements on an Eastern Indonesian Island* (Duke University, 2000).
Statistical Pocketbook of Bangladesh 1994 (Dhaka: Bangladesh Bureau of Statistics, 1995).
Stoler, Ann Laura, and Karen Strassler, 'Memory Work and Colonial Studies: Recasting the Colonial in Contemporary Java,' paper presented at the Research Centre Religion and Society, Amsterdam, April 1998.
Subandi, Ketut, *Berbakti Kepada Kawitan (Leluhur) Adalah Paramo Dharmah* (Denpasar, 1982).
Suchit Wongthet (ed.), Journal *Sinlapawathanatham* (Special Issue, 1984).
Swellengrebel, J. (ed.), *Bali, Studies in Life, Thought and Ritual* (The Hague: Van Hoeve, 1960).
Tanabe, Shigeharu, 'Historical Geography of the Canal System in the Chao Phraya Delta,' *Journal of the Siam Society*, 65:2 (July 1977), 23-72.
Terwiel, B.J., 'Foreword,' in W.C. Dodd, 1996.
Thailand, Kingdom of, Office of the Prime Minister, *Thailand into the 1980s* (Bangkok: Office of the Prime Minister, 1979).
Thi raluek Sayamratthaphiphithaphan suan Lumphini Phraphutthasakkarat 2468 (The Souvenir of the Siamese Kingdom Exhibition at Lumbini Park, BE 2468).
Thompson, E.P., 'Time, Work-Discipline, and Industrial Capitalism,' *Past and Present*, 38 (1967), 57-59.
Toma Seita, *Nihon minzoku no keisei* (Tokyo: n.p., 1951).
Tôyama Shigeki, *Sengo no rekishigaku to rekishi ishiki* (Tokyo: n.p., 1968).
Tôyama Shigeki, Imai Seiichi, and Fujiwara Akira, *Shôwashi* (Tokyo: n.p., 1955).
Tôyama Shigeki, Imai Seiichi, and Fujiwara Akira, *Shôwashi. Shinpan* (Tokyo: n.p., 1959).
Tôyama Shigeki and Nagahara Keiji, 'Jidai kubun ron,' *Nihon rekishi*, bekkan 1, Iwanami Kôza (Tokyo: n.p., 1963), 167-226.
Trautmann, Thomas R., 'Indian Time, European Time,' in Hughes and Trautmann (eds.), 1995, 167-197.
Turner, Bryan S., *Weber and Islam: A Critical Study* (London: n.p., 1974).
Turner, Bryan S., *Marx and the End of Orientalism* (London: n.p., 1978).
Ubukata Naokichi, Tôyama Shigeki, and Tanaka Masatoshi (eds.), *Rekishizô saikôsei no kadai: Rekishigaku no hôhô to Ajia* (Tokyo: n.p., 1966).
Umesao Tadao, 'Bunmei no seitai shikan josetsu,' *Chûô kôron*, Feb. 1957, 32-49.

Van der Veer, Peter (ed.), *Conversion to Modernities: The Globalization of Christianity* (New York: Routledge, 1996)
Van Eijbergen, H.C., 'Verslag eener reis naar de Aroe- en Key Eilanden in de maand Junij, 1862,' *Tijdschrift voor Indischen Taal-, Land- en Volkenkunde*, 14 (1866), 220-272.
Van Höevell, G.W.W.C. Baron, 'De Aroe-Eilanden, Geographisch, Ethnographisch en Commercieel,' *Tijdschrift voor Taal-, Land- en Volkenkunde*, 32 (1890), 1-45.
Van Leur, J.C., *Indonesian Trade and Society* (The Hague: W. van Hoeve, Ltd., 1983).
Van Schendel, Willem, *Reviving a Rural Industry: Silk Producers and Officials in India and Bangladesh 1880s to 1980s* (Dhaka: University Press Limited, 1995).
Van Schendel, Willem, 'Jatir Hoye Ke Bole? Jatiyotabadi Shunyogorbho Boktrita Ebong Shangskritik Bohuttobader Protibad,' in Willem van Schendel and Ellen Bal (eds.), 1998, 222-263. Also forthcoming as 'Who Speaks for the Nation? Nationalist Rhetoric and the Challenge of Cultural Pluralism in Bangladesh' in Willem van Schendel and Erik-Jan Zürcher (eds.), 2001.
Van Schendel, Willem, Wolfgang Mey, and Aditya K. Dewan, *The Chittagong Hill Tracts: Living in a Borderland* (Bangkok: White Lotus, 2000).
Van Schendel, Willem, and Ellen Bal (eds.), *Banglar Bohujati: Bangali Chhara Onnanyojatir Proshongo* (Calcutta: International Centre for Bengal Studies, 1998)
Van Schendel, Willem, and Erik-Jan Zürcher (eds.), *Identity Politics in Central Asia and the Muslim World* (London: I.B. Tauris, 2001).
Vella, Walter F., *Chaiyo! King Vajiravudh and the Development of Thai Nationalism* (Honolulu: The University Press of of Hawaii, 1978).
Vickers, A., 'The Historiography of Balinese Texts,' *History and Theory*, 29:2 (1990), 158-178.
Vickery, Michael, 'A Guide through some Recent Sukhothai Historiography,' *Journal of the Siam Society*, 66 (July 1978), 182-246.
Vickery, Michael, 'A New Tamnan about Ayudhya,' *Journal of the Siam Society*, 67:2 (1979), 123-186.
Warren, C., *Adat and Dinas: Balinese Communities in the Indonesian State* (Kuala Lumpur: Oxford University Press, 1993).
Warren, C., 'Centre and Periphery in Indonesia: Environment, Politics and Human Rights in the Regional Press (Bali),' Working Paper 42 (Perth: Asia Research Centre on Social, Political and Economic Change, Murdoch University, 1994).
Wehler, Hans-Ulrich, *Modernisierungstheorie und Geschichte* (Göttingen: n.p., 1975).
Wehler, Hans-Ulrich, 'Deutscher Sonderweg' oder allgemeine Probleme des westlichen Kapitalismus?,' *Merkur*, 35 (1981), 478-487.
Wessels, C., 'Een Portugese Missiepoging op Bali in 1635,' *Studiën*, 55 (1923), 433-443.
Wood, Geoffrey D., *Bangladesh: Whose Ideas, Whose Interests?* (Dhaka: University Press Limited, 1994).
Yamada Mitsuhiro, 'Sengo 'Nihon tsûshi' bunken mokuroku 1945-85,' *Rekishi hyôron*, 554 (1996), 81-86.
Yamamoto, Tatsuro, 'Thailand as it is Referred to in the Da-de Nan-hai zhi at the Beginning of the Fourteenth Century,' *Journal of East-West Maritime Relations*, 1 (1989), 47-58.

Yoon Keun-Cha, 'Sengo rekishigaku ni okeru tasha ninshiki: Zainichi Chôsenjin no shiten kara,' *Rekishigaku kenkyû*, 594 (1989), 2-16.

Yoon Keun-Cha, 'Sengo rekishigaku no Ajiakan,' in Asao Naohiro, Kano Masanao et al., (eds.) 1995, 249-279.

Young, Michael, *The Metronomic Society: Natural Rhythms and Human Timetables* (Cambridge, Mass.: Harvard University Press, 1988).

Zbavitel, Dusan, *Bengali Folk-Ballads from Mymensingh and the Problem of their Authenticity* (Calcutta: University of Calcutta, 1963).

Zerubavel, Eviatar, *Hidden Rhythms: Schedules and Calendars in Social Life* (Berkeley and Los Angeles: University of California Press, 1981).

Zerubavel, Eviatar, 'The Standardization of Time: A Sociohistorical Perspective,' *American Journal of Sociology*, 88 (1982).

Zerubavel, Eviatar, *The Seven Day Circle: The History of Meaning of the Week* (New York: Free Press, 1985).

Zunz, Olivier (ed.), *Reliving the Past: The Worlds of Social History* (Chapel Hill and London: n.p., 1985).

Contributors

Sebastian Conrad
Arbeitsstelle für Vergleichende Gesellschaftsgeschichte, Freie Universität Berlin

Johan Goudsblom
Emeritus Professor of Sociology, University of Amsterdam

Henk Schulte Nordholt
Professor of Asian History, Erasmus University Rotterdam / Asian Studies, IIAS, University of Amsterdam

Patricia Spyer
Professor of Sociology and Anthropology of Indonesia, University of Leiden

Barend J. Terwiel
Professor of Thai Studies, Abteilung Thai- und Vietnamstudien, Universität Hamburg / IIAS Extraordinary Chair for Mainland Southeast Asia, University of Leiden

Willem van Schendel
Professor of Modern Asian History, University of Amsterdam / International Institute of Social History, Amsterdam

COMPARATIVE ASIAN STUDIES

General editor: Leontine Visser

PUBLICATIONS IN THIS SERIES

1 CONCEPTUALIZING DEVELOPMENT – THE HISTORICAL-SOCIOLOGICAL TRADITION IN DUTCH NON-WESTERN SOCIOLOGY / Otto van den Muyzenberg and Willem Wolters / isbn 90-6765-382-9 39 p.
2 THE SHATTERED IMAGE – CONSTRUCTION AND DECONSTRUCTION OF THE VILLAGE IN COLONIAL ASIA / Jan Breman / isbn 90-6765-383-7 50 p.
3 SEDUCTIVE MIRAGE – THE SEARCH FOR THE VILLAGE COMMUNITY IN SOUTHEAST ASIA / Jeremy Kemp / isbn 90-6765-384-5 47 p.
4 BETWEEN SOVEREIGN DOMAIN AND SERVILE TENURE – THE DEVELOPMENT OF RIGHTS TO LAND IN JAVA, 1780-1870 / Peter Boomgaard / isbn 90-6765-788-6 61 p.
5 LABOUR MIGRATION AND RURAL TRANSFORMATION IN COLONIAL ASIA / Jan Breman / isbn 90-6765-873-4 82 p.
6 LIVING IN DELI: ITS SOCIETY AS IMAGED IN COLONIAL FICTION / Lily E. Clerkx and Wim F. Wertheim / isbn 90-6765-965-X 126 p.
7 STATE, VILLAGE AND RITUAL IN BALI – A HISTORICAL PERSPECTIVE / Henk Schulte Nordholt / isbn 90-5383-023-5 58 p.
8 THE CENTRALITY OF CENTRAL ASIA / Andre Gunder Frank / isbn 90-5383-079-0 68 p.
9 IDEOLOGICAL INNOVATION UNDER MONARCHY – ASPECTS OF LEGITIMATION ACTIVITY IN CONTEMPORARY BRUNEI / G. Braighlinn / isbn 90-5383-091-X 112 p.
10 THE STATE OF BIHAR / Arvind N. Das / isbn 90-5383-135-5 116 p.
11 ON THE PRODUCTION OF KNOWLEDGE – FIELDWORK IN SOUTH GUJARAT, 1971-1990 / Hein Streefkerk / isbn 90-5383-188-6 54 p.
12 COMPARATIVE ESSAYS ON ASIA AND THE WEST / Wim F. Wertheim / isbn 90-5383-196-7 109 p.
13 COLONIAL PRODUCTION IN PROVINCIAL JAVA – THE SUGAR INDUSTRY IN PEKALONGAN-TEGAL, 1800-1942 / G.R. Knight / isbn 90-5383-260-2 76 p.
14 ASIAN CAPITALISTS IN THE EUROPEAN MIRROR / Mario Rutten / isbn 90-5383-270-X 72 p.
15 A PEOPLE OF MIGRANTS – ETHNICITY, STATE AND RELIGION IN KARACHI / Oskar Verkaaik / isbn 90-5383-339-0 89 p.
16 COMMUNITIES AND ELECTORATES – A COMPARATIVE DISCUSSION OF COMMUNALISM IN COLONIAL INDIA / Dick Kooiman / isbn 90-5383-394-3 88 p.
17 SOCIAL SCIENCE IN SOUTHEAST ASIA – FROM PARTICULARISM TO UNIVERSALISM / Nico Schulte Nordholt and Leontine Visser (eds.) / isbn 90-5383-427-3 165 p.
18 MARTYRDOM AND POLITICAL RESISTANCE – ESSAYS FROM ASIA AND EUROPE / Joyce Pettigrew (ed.) / isbn 90-5383-501-6 146 p.
19 UNSETTLED FRONTIERS AND TRANSNATIONAL LINKAGES – NEW TASKS FOR THE HISTORIAN OF MODERN ASIA / Leo Douw (ed.) / isbn 90-5383-539-6 38 p.
20 THE SULU ZONE – THE WORLD CAPITALIST ECONOMY AND THE HISTORICAL IMAGINATION / James Francis Warren / isbn 90-5383-568-7 71 p.